Social Value in Constr

T0259051

While the concept of social value is not new, recent interest in social value in construction has grown because of new social procurement legislation around the world and an increasing acceptance of the need to ensure construction projects provide social value, rather than simply economic value. Despite this growing recognition, literature and professional guidance on the subject is hard to find. This is the first book looking at social value in construction and it sets the agenda by asking and answering important questions like:

- How is the construction industry developing and supporting social enterprise and social value and for who?
- How and when is the industry recording and measuring social value and its effect?
- Which organisations are doing things well and what can we learn from their experiences?
- What can industry players do together to consolidate efforts and drive improvements?
- What are the key challenges in the field and what does the future look like?

Drawing on a variety of professional and academic experiences and disciplines, the authors present global perspectives and lay the foundations for creating social value in the construction industry. This timely book makes use of real-life case studies and examples of best practice to demonstrate how innovative companies can utilise contemporary research to create social value through their projects. It is time the construction industry viewed community involvement and corporate social responsibility as an opportunity rather than a risk, and this is the book that shows the industry how. This is essential reading for all professionals in the construction, engineering, architecture and built environment sector. In particular, project managers, clients, contract managers, quantity surveyors, CSR and HR personnel will gain a lot from reading this book.

Ani Raiden is Senior Lecturer in Human Resource Management at Nottingham Business School, Nottingham Trent University, UK. She is the Immediate Past Chair of the Association of Researchers in Construction Management, Fellow of the Higher Education Academy, and Chartered Member of the Chartered Institute of Personnel and Development.

Martin Loosemore is Professor of Construction Management at the University of New South Wales, Sydney, Australia. He is a Fellow of the Chartered Institute of Building and has published many books and articles in the areas of social enterprise and procurement, innovation, risk management, corporate social responsibility and corporate strategy.

Andrew King runs Soul Value, a built environment consultancy based in the UK, where he draws on both his extensive construction background and his work in the field of honest and authentic communication to help clients innovatively maximise social value on their projects.

Chris Gorse is Professor of Construction and Project Management, the Director of the Leeds Sustainability Institute, and Head of the Centre for the Built Environment at Leeds Beckett University, UK.

Social Value in Construction

Ani Raiden, Martin Loosemore,
Andrew King and Chris Gorse

LONDON AND NEW YORK

First published 2019
by Routledge
2 Park Square, Milton Park, Abingdon, Oxon OX14 4RN

and by Routledge
52 Vanderbilt Avenue, New York, NY 10017

Routledge is an imprint of the Taylor & Francis Group, an informa business

British Library Cataloguing-in-Publication Data
A catalogue record for this book is available from the British Library

Library of Congress Cataloging-in-Publication Data
A catalog record has been requested for this book

ISBN: 978-1-138-29509-4 (hbk)
ISBN: 978-1-138-29510-0 (pbk)
ISBN: 978-1-315-10080-7 (ebk)

Typeset in Times
by Out of House Publishing

Contents

List of illustrations

Figures

Tables

About the authors

Ani Raiden, Nottingham Business School, Nottingham Trent University, UK
Ani Raiden is Senior Lecturer in Human Resource Management at Nottingham Business School, Nottingham Trent University, UK. She is Immediate Past Chair of the Association of Researchers in Construction Management, Fellow of the Higher Education Academy, and Chartered Member of the Chartered Institute of Personnel and Development. Ani studied for her PhD at the Department of Civil and Building Engineering at Loughborough University in the UK, developing a strategic employee resourcing framework for construction organisations. This bridged her academic interests and experience in the industry. As a human resource transformation executive (Diversity) at Eircom, an Irish communications company, Ani specialised in equal opportunities and managing diversity before embarking on an academic career. Ani's research builds on her work on people resourcing and focuses on 'quality of working life'. She is a co-author of *Employee Resourcing in the Construction Industry*, and has published many articles in the areas of work-life balance, well-being, health and safety, knowledge management, and human resource management and human resource development in construction.

Martin Loosemore, University of New South Wales, Australia Martin Loosemore is Professor of Construction Management at the University of New South Wales, Sydney, Australia. He is a Fellow of the Chartered Institute of Building and has published many books and articles in the areas of social enterprise and procurement, innovation, risk management, corporate social responsibility and corporate strategy. His work has received numerous international awards and he serves on many international advisory committees and editorial boards. Martin was appointed advisor on international workplace productivity and reform to the Australian Federal Government's 2003 Royal Commission into the building and construction industry. He was subsequently called to provide evidence to the 2004 Federal Senate inquiry into the Building and Construction Industry Bill and the 2009 Federal Senate inquiry into the Building and Construction Industry Improvement Amendment (Transition to Fair Work) Bill 2009. In 2009 he was appointed to the Australian Government's Built Environment Industry Innovation Council which advised the Federal

Minister for Innovation, Industry Science and Research on innovation policy in areas like climate change, sustainability and industry competitiveness as well as issues such as regulatory reform, workforce capability, skills needs, new technologies and other priorities for the industry. In 2014 Martin contributed significantly to the 2014 Productivity Commission Inquiry report into infrastructure and he serves on the Western Sydney Social Procurement and Jobs Opportunities Steering Committee, New South Wales Government, Department of Premier and Cabinet. Martin is the founding partner of a successful social business which specialises in securing employment opportunities in the construction and engineering industries for people suffering disadvantage.

Andrew King, Soul Value Andrew King runs Soul Value, a built environment consultancy based in the UK, where he draws on both his extensive construction background and his work in the field of honest and authentic communication to help clients innovatively maximise social value on their projects. Soul Value takes a participatory and collaborative approach and is founded on the positive and transformative impact of people getting out of their heads, into their bodies, and relating with honesty and simplicity. Andrew previously worked in the construction industry where he ran his own surveying business, practised as a quantity surveyor, and eventually played a central role in developing a relationship-based supply chain strategy for one of the UK's largest contractors, now Morgan Sindall. He then took a senior lecturer position in the School of Architecture, Design and the Built Environment at Nottingham Trent University with a mission to help students gain a real-life understanding of the industry. His current focus on social value is a progression of his earlier value-based research highlighting the importance of transitioning client value through the wider supply chain by focusing on main contractor-subcontractor tender processes, itself the focus of his PhD. Andrew is a fellow of the Higher Education Academy and has published a range of outputs in design and build tendering, supply chain management and computer aided qualitative data analysis software. The Chartered Institute of Building recognised Andrew's involvement in an Engineering and Physical Sciences Research Council research project seeking to improve tender evaluation in design and build projects by jointly awarding him the Premier Research and Innovation Award.

Chris Gorse, Leeds Beckett University, UK Chris Gorse is Professor of Construction and Project Management, the Director of the Leeds Sustainability Institute, and Head of the Centre for the Built Environment at Leeds Beckett University, UK. He is an Engineering Professors' Council member, member of the Chartered Institute of Building, member of the Association for Project Management, Fellow of the Higher Education Academy, and the chair of the Association of Researchers in Construction Management. He sits on the board of a number of large education academies in the UK and has a keen interest in improving opportunities and education for all. Chris worked as an engineer and legal consultant before pursuing a career in education. He has a degree in building engineering, a masters in construction law, and a PhD in applied psychology and group behaviour; and he is an established author with publications in communication, technology, law, energy and management. Chris has provided evidence to all party enquiries held at the Houses of Parliament on the quality of building in the UK. The work of his research group at Leeds Sustainability Institute is diverse, with physicists, data analyst and behavioral scientists all focused on understanding the various sociotechnical implications of the built environment. The group's work on energy and buildings has resulted in revisions to the building regulations in the UK, making significant interventions towards reducing energy consumption in buildings. Chris is the principle investigator on a number of built environment projects, which are varied in nature but all focused on improving the built environment. Fields of enquiry include payment schemes and social value frameworks; regulation and management; smart meter data and social implications; health, activity and its relationship with the built environment; and fuel poverty.

Beulah Allaway and Martin Brown, Anthony Collins Solicitors LLP, UK
Beulah is an associate in the construction team at Anthony Collins Solicitors, UK, specialising in construction and procurement advice to housing, local government, health and social care, education and social business sectors. Beulah has a longstanding interest in social value in public procurement and advises clients on maximising social value through public works procurement.

After pursuing a training contract with Anthony Collins Solicitors, Martin qualified as a solicitor in September 2018. Martin's interest in social value stems from his master's degree obtained at the University of Law, in which he focused specifically on the relationship between social value and the wider public procurement rules.

Both are part of a wider social value team at Anthony Collins Solicitors that includes Senior Associate Gayle Monk and is led by partner Mark Cook. As a team, they provide advice on all aspects of achieving social value through public contracts, supported Chris White MP in his promotion of the Public Services (Social Value) Act 2012 and have produced some leading guidance on social value, such as 'Social value and public procurement: a legal guide', which was supplementary to Richard Macfarlane's 2014 report 'Tackling poverty through public procurement' for the Joseph Rowntree Foundation.

Kate Canning, Arup Kate Canning is Associate Geo-Environmental Engineer with Arup specialising in the design and delivery of brownfield remediation schemes. On completion of her PhD, Kate was a research fellow in the Centre for Contaminated Land Remediation, where she led an industrial research project funded by Blue Circle Cement. Since joining Arup, Kate has been actively involved in the promotion of Arup's contaminated land business and skills network across the UK. Subsequently she helped form the Northwest Brownfield Regeneration Forum, and sits on the steering committee. More recently, Kate has been focusing on the site restoration challenges associated with the decommissioning of the UK's nuclear estate and the strategic commercial and non-commercial use of land assets. Kate has been Arup's Research Leader for the UK, Middle East and Africa since 2013, and is responsible for managing UKMEA strategic investment funds and strategic relationships with universities and research councils.

Paul W. Chan, University of Manchester, UK Paul W. Chan is Senior Lecturer and Lead of the Management of Projects Research Group at The University of Manchester, UK. He has a track record in studying human relations in engineering and construction projects, with particular emphasis on how people cope with social, organisational and technological change. He manages a number of industry-funded PhD projects that examine business model innovation and engagement across the value chain. He is also collaborating with Arup and the UK Nuclear Decommissioning Authority on a PhD project on social value communication in decommissioning and site remediation. He has co-authored a book on leadership in construction entitled Constructing Futures with Professor Rachel Cooper (Wiley-Blackwell, 2010), and published over 80 peer-reviewed journal and conference papers. He is an editor of *Construction Management and Economics* and was Chair of the Association of Researchers in Construction Management.

Sophia de Sousa, Chief Executive, The Glass-House Community Led Design, UK Sophia joined The Glass-House Community Led Design, UK, as Chief Executive in 2005. She is an impassioned champion of design quality and enabler of design practices that empower people and organisations and that help communities thrive. She is also a leader in the field of research on community-led, participatory and co-design practices and is a visiting fellow at the Open University. Sophia plays an active role in designing and delivering hands-on training and support to communities and practitioners, and in co-designing research, innovation and resources. She is also an active spokesperson and expert advisor, contributing regularly to events and to panels and advisory groups including: the Historic England Urban Panel; the Oxford Design Review Panel; the Barking and Dagenham Design Advice Panel; the Highways England Design Review Panel; Design Council Cabe Built Environment Experts; and the Historic England Expert Advisory Group. Sophia's background is in architecture and urbanism, education, voluntary sector work and multiculturalism. She is passionate about finding opportunity in complexity and diversity, and is committed to working collaboratively to finding informed and creative solutions. Sophia is fascinated by the connection between design, people and place, and believes that our sensory and emotional connections with the places around us shape how we live, and who we are.

Dave Higgon, Multiplex, Australia Dave Higgon is Employee Relations Manager for Multiplex Constructions, Australia. He is a co-author of *Risk Management in Projects, and Social Enterprises in Construction*. His background includes time spent as a tradesperson, a subcontractor, a building union official, a safety manager, a consultant and currently the employee relations manager for Multiplex Constructions. His extensive and practical industry experience has contributed to a range of innovative and successful solutions in the areas of industrial relations, occupational health and safety, and training and development. Key themes in these solutions have been collaboration and the promotion of cooperative relationships between industry stakeholders most recently in regard to developing employment opportunities in the building and construction industry for groups suffering various forms of disadvantage.

Andrew Knight, Nottingham Trent University Andrew Knight is Dean of the School of Architecture, Design and the Built Environment at Nottingham Trent University, UK. This includes the strategic leadership of a school of 3,000 students covering disciplines from civil engineering to product design. Andrew commenced his career in local government as a trainee surveyor before moving to private practice. In the late 1990s he was appointed as a lecturer at Sheffield Hallam University. He then moved to Nottingham Trent University in 2003 as a principal lecturer before moving to the role of Head of Department in 2008. After seven years of successful departmental management, overseeing a range of strategic initiatives and changes, he was appointed to a number of senior positions before accepting the role of Dean of School in 2018. Andrew continues to teach on various courses across the school and has a particular interest in philosophical ideas applied to professional contexts; for example, professional ethics, historic architecture, aesthetics and education. He has successfully supervised over ten PhD students and is on the editorial board of the Innovation in the Built Environment book series, published by the RICS and Wiley-Blackwell. Additionally, he sits on a variety of external committees. He is a qualified chartered surveyor, a philosophy graduate and he holds a PhD from the University of Nottingham.

Anna Mimms, MBE Anna Mimms, MBE, is a senior strategist, consultant, and a public speaker with a developing reputation in the field of social return on investment in the UK, which was cited on her appointment to MBE in 2013. Anna works across a number of sectors as an expert in change and transition, specialising in complex and challenging business environments. Her work in social value has seen her found an award winning social enterprise, deliver tangible social impact on a number of construction projects, and influence public sector procurement practices to include meaningful social causes. In 2014 the Construction Industry Training Board recognised Anna's excellence in embedding training as a priority in a social enterprise construction company, and employing and developing apprentices and employees from complex backgrounds. In the same year, the Chartered Institute of

Building commended her contributions towards developing meaningful private, public, third sector partnerships in the construction industry. This work has been a response to specific economic hardships associated with complex social environments. Anna has also been an active volunteer supporting the Dyslexia Association, Framework, Nottingham Women's Network, and the WellWomen Centre in Wakefield.

Cara Mulholland, University of Manchester, UK Cara Mulholland is a PhD researcher in the Management of Projects Research Group within the School of Mechanical, Aerospace and Civil Engineering at the University of Manchester, UK. Her PhD is funded through the Engineering and Physical Sciences Research Council as an industry collaboration project with Arup and the UK's Nuclear Decommissioning Authority. She researches social value communication in practice in large scale engineering projects, primarily in the nuclear decommissioning sector. Her research explores how the perceived, measured and reported social value may change within the framings of time, scope and place. Her interest in the social value of the built environment grew from seeing the connection between people and technology during her undergraduate degree, and then studying towards a masters in civil engineering (with enterprise), where she championed engineers' responsibility to the triple bottom line of sustainability through various outreach programmes and charity work.

Joanne Osborne, Damajo Consulting Pty Ltd, Australia Joanne Osborne, is a cofounder and principle consultant of Damajo Consulting, Australia, specialising in social procurement, risk and diversity. Jo has over two decades experience working in Community and Charitable organisations and has been responsible for the development of many successful models that have been applied in projects for not-for-profit organisations, government and the private sector around Australia. In 2012 Jo received the inaugural award Services of Excellence to the National Employer Services Industry from the National Employer Services Association. Over the past seven years Jo has worked closely with Multiplex in the implementation of their LINC (Linking Industry Needs with Community) programme, she lectures at the University of New South Wales, and for three years held the role of Executive Manager for The Literacy For Life Foundation. Her current role involves working with Multiplex in managing the Connectivity Centre Programmes.

Preface

Social value is concerned with how we contribute positively to the communities in which we work. The purpose of this book is to explore the emerging concept of social value in the context of the built environment. By this we mean every professional and firm involved with the entire life-cycle of construction from planning through design, construction, operations and facilities management. Our aim is to create a conceptual and practical foundation to advance thinking and practice in this new and rapidly emerging field and to show that there are many different motives, rationales and methods for creating social value in practice. There are many untapped opportunities to grasp and challenges to overcome in achieving this important goal and we recognise that we are at the beginning of an ongoing and interesting journey which will draw knowledge from many disciplines and fields of knowledge. No one organisation or field of knowledge can alone solve the complex social challenges we face, and it is clear that solutions will need to be co-created through cross-sector collaboration between a wide range of stakeholders. In this book we discuss the current state of development in the field of social value and reflect on how it applies to the built environment. In order to help the built environment make a more positive impact on the communities in which it builds, our aims are simple: to contribute towards the development of a common language that aids communication of the key concepts; the identification, through practical examples from thought-leaders in the field, of what does and doesn't work; and the mapping of a clear agenda and a plan to advance thinking in this field.

We are optimistic since health and safety is a parallel journey which illustrates how the built environment can change its culture. Whilst construction sites are still unacceptably dangerous workplaces, the advances in the industry's safety record in recent years are the result of sustained effort combining legislative measures; research, education and training; and leadership. We believe that the same shifts can be achieved in the field of social value. We want to dismantle the physical and psychological hoardings we build around our projects and educate our professionals to treat the community as an asset rather than a risk. And we also want to better measure and acknowledge the enormous actual and potential value we can contribute to society, much of which we feel remains untapped.

To this end, we integrate theory and practice into a value-driven approach to social value, where creating and managing social value becomes an integral part of business decision-making, organisation and practice. We discuss strategy, policy, and operational issues, and highlight good practice. We draw on current conceptual and practical knowledge from within, and importantly beyond, the built environment, bringing in research and practice from the wider field of business and management. We also draw on the behavioural approach, looking to psychology and sociology, which allow us to develop discussion about social value orientation, and how that may influence the choices individuals make, and hence how managers, clients and other stakeholders involved with construction projects may think about social value.

We understand that our readership, different people and different organisations, will be at different places along the journey; some fully embracing the concept of social value, some only thinking about compliance with legislation. We are offering a knowledge and resource bank, and we believe that this book will have some new insights and lessons for all. We take a bottom-up approach based on the belief that it is society, communities and individuals that determine the extent and nature of social value that is created through construction projects; rather than what so-called experts, from outside those communities think is best. In addition, a cautionary approach to using complex models that create distance from, rather than closeness to, the values and outcomes they try to model and measure is advised. In this way, creating social value needs to be firmly founded on a deep and pragmatic recognition of what value means to the communities impacted; a message we believe is fundamental and continue to make throughout the work.

The first of its kind on social value in the built environment, this book is organised in two parts:

Part 1 lays out the principles and conceptual foundations to social value.
Part 2 presents a range of good practice case studies.

We hope you find this book of value and we hope it makes a difference. Social Value is topical today, connecting closely with the United Nations' Global Sustainable Development Goals, which are a universal call to action to end poverty, protect the planet and ensure that all people enjoy peace and prosperity. The Sustainable Development Goals work in the spirit of partnership and pragmatism at a national level encouraging countries to make the right choices now to improve life, in a sustainable way, for future generations. Social Value is a vehicle and a goal for organisations and projects to make a positive contribution now and over longer-term.

Ani Raiden, Martin Loosemore, Andrew King and Chris Gorse

Acknowledgements

Research on social value in the built environment is increasing in momentum and we are excited that the development of this book is in part the result of the following events. We would like to thank the organisations and individuals who have supported this book.

First, we thank the Association of Researchers in Construction Management (ARCOM) Committee for financial support and 'air time'. The conversation began at an ARCOM sponsored Social Value Seminar held at Nottingham Business School, Nottingham Trent University, in the UK in March 2016. A small group of invited participants represented the major players in this space, including private and public sector contractors and clients, social enterprises, researchers and academics. The aim of the day was to draw together some of the leading minds in social enterprise and social value in the built environment to better define the field of enquiry, share experiences and consolidate existing efforts. Presentations by Beulah Allaway, Anthony Collins Solicitors; Dan Chaplin, Latch; Ian Drayton, Soarbuild; Paul Senior, Keepmoat; Simon Gutteridge, Wakefield District Housing; and Anna Mimms, MBE, helped crystallise the need for a text that is accessible for both researchers and practitioners.

Many of the seminar presenters have contributed to the development of this book, and we are grateful for their ongoing commitment and collaboration with the project. Specifically, thanks to Beulah and her colleague Martin for writing Chapter 2 on law; and to Anna for the section she has co-authored with us in addition to the many meetings and soul searching discussions we have had over the past year.

In 2016 we also hosted a Social Value Spotlight session within the 32nd annual ARCOM conference in Manchester, UK. Presentations by Sylvia Hammond, University of Cape Town, South Africa; Daniella Petersen, Chalmers University of Technology, Sweden; Fady Farag, University of Salford, UK; and Jemma Bridgeman, Construction Youth Trust, Wales, UK, stimulated much discussion and debate. We thank all the presenters and other conference participants who came and contributed to this session. It was after the Social Value Spotlight that we agreed on our interests and intentions to write this text and met with Ed Needle from Taylor & Francis to discuss our initial ideas. His enthusiasm and editorial guidance throughout the project have played a central supporting role in successfully bringing the book together.

We would also wish to thank the reviewers of the first proposal for this book who gave us much supportive and constructive feedback – this helped in growing our ideas.

One of the case studies which we present in Part 2 of the book, the British Land redevelopment project at Meadowhall Shopping Centre, offered a very valuable opportunity to explore social value from multiple perspectives. British Land's significant experience in the creation and management of social value allied to Laing O'Rourke's important and sustained assistance and access to a wide range of team members, including their supply chain, helped sensitise and deepen the case study but also contributed to the development of the book as a whole. Lesley Giddins of Sandgrown Consulting, working as British Land's social value consultant on the Meadowhall project, deserves special mention for the value her depth and breadth of experience brought to the work.

We also like to thank Louise Townsend, Head of Social Value at Morgan Sindall, whose expert counsel gained through years of experience driving innovation in social value helped refine this work.

Last, but absolutely by no means least, we thank the other authors who have provided contributions to the text. Andrew Knight, Nottingham Trent University, brings a philosophical dimension to our work that develops a moral imperative for social value. Sophia de Sousa, Glass-House Community Led Design; Dave Higgon, Multiplex; Joanne Osborne, Damajo Consulting Pty Ltd; Cara Mulholland and Paul W Chan, The University of Manchester; and Kate Canning, Arup – your case studies have brought social value to life and show the path forward by documenting innovative and practical ways for the built environment globally to create social value in partnership with the communities in which we build.

In the spirit of co-creating social value, this book is a truly collaborative achievement. Thank you.

Part 1

Principles and conceptual foundations

Part 1 of our book lays out the principles and conceptual foundations to social value. In the introduction we define social value and discuss the language of social value. We then consider the many drivers for thinking about and creating social value, together with the political context of social value. We begin to map out some practical examples of what social value may mean in the built environment, and highlight specific projects that have evidenced social value in practice throughout the construction project's life-cycle from planning and design, to tendering, construction and facilities management. Allaway and Brown then present a review of the legal framework and associated practical implications in the UK in Chapter 2. This is followed by a theoretical justification for social value by Knight in Chapter 3. In Chapter 4 we discuss different ways of creating social value within and between organisations, through strategic initiatives, such as partnering with social enterprises, and some specific ways of achieving impact through employment and training. Embedding social value to strategy and practice arises as a key theme in our discussion. Finally, Chapter 5 outlines current debates and different models and tools regarding the assessment and measurement of social value.

1   Introduction

*Ani Raiden, Martin Loosemore, Andrew King
and Chris Gorse*

*In this chapter we aim to define social value and demystify the somewhat con-
fusing language that is used in this area. Through reference to contemporary
and numerous practical examples, we also discuss emerging trends and drivers
of social value in the built environment and the ways in which social value can
be created throughout a construction project's life-cycle from planning and
design, through tendering, construction and facilities management. The chapter
concludes with a discussion about the ethics of social value in the built environment and raises important questions for practitioners, students and researchers
about the many perspectives that must be considered when developing, measuring
and reporting the social impact initiatives that are designed to give back to the
communities in which the industry builds.*

Social value in context

The built environment has a major impact on the communities in which it builds,
although the social dimensions of this impact have been relatively neglected in
comparison to the economic and ecological (environmental) issues. In academic
circles, the recent interest in the concept of 'social value' in the built environment has been largely a response to the need to fill this gap in our knowledge. In
practice, social value has been driven by legislation, procurement, and growing
community demands for greater corporate social responsibility (CSR) in built
environment business practices. The World Business Council for Sustainable
Development defines CSR as 'the continuing commitment by business to behave
ethically and contribute to economic development while improving the quality
of life of the workforce and their families as well as of the local community and
society at large' (Watts and Holme 2003: 3).

In the context of construction, CSR is about the relationship between
organisations and society and as Watts, Dainty and Fernie (2015) point out, a
good CSR record is becoming an increasingly important selection criteria for
both public and private construction industry clients employing firms within the
built environment. Social procurement is just one increasingly popular mechanism by which construction clients are enforcing their CSR expectations on the
built environment, which in simple terms involves leveraging their purchasing

power with the deliberate aim of creating social value in the communities in which they build (Barraket, Keast and Furneaux 2016, Barraket and Loosemore 2018. While there remains a 'lack of construct clarity'around the emerging concept of social procurement (Furneaux and Barraket 2014: 265) social procurement is broadly defined as 'the acquisition of a range of assets and services, with the aim of intentionally creating social outcomes (both directly and indirectly)' (ibid: 269). In simple terms, 'direct' social procurement involves purchasing construction products and services directly from socially responsible businesses and social benefit organisations which trade for a social purpose. These would include, for example, social enterprises, indigenous businesses, enterprises owned by people with disabilities, minority owned enterprises, enterprising not-for-profits/ charities, social businesses, cooperatives, enterprising charities and local businesses. In contrast, 'indirect' social procurement involves requiring business partners in existing construction supply chains to do the same, through numerous mechanisms such as social clauses in employment contracts; supplier codes of practice; responsible sourcing policies. Through social procurement initiatives, businesses diversify their project supply chains with the dual goal of maximising both economic and social value for their shareholders, stakeholders and clients. This social value can take many forms. For example, some social procurement initiatives may be targeted at employing local businesses while others may be aimed at providing employment and training opportunities to disadvantaged groups such as the long-term unemployed, disengaged youth, ex-offenders, people with disabilities or indigenous groups. These outcomes can in-turn translate to numerous impacts for wider society such as improved income, health and well-being and reduced crime, substance abuse and incarceration, which social impact practitioners controversially attempt to measure, quantify and monetise using a variety of techniques such as social return on investment (SROI) (Maier et al. 2015).

As Loosemore and Higgon (2015) and Petersen and Kadefors (2016) point out, due to its large size and its potential multiplier effect into the wider economy, the built environment is increasingly seen by governments (and major socially responsible private clients) as a powerful tool to tackle complex social problems which seem resistant to traditional government welfare interventions and policies. In line with emerging and contemporary principles of 'New Public Governance', one of the defining features of social procurement is that rather than governments working alone to tackle social problems like entrenched unemployment, social problems are resolved through new cross-sector collaborations and partnerships between the government, private and third sectors (Furneaux and Barraket 2014; Barraket, Keast and Furneaux 2016). As we will reveal, these new expectations to create social value by collaborating across previously disconnected sectors, raise many new challenges, but also many opportunities, for professionals and businesses operating in the built environment.

One of the immediate challenges, both within and outside the construction sector, is that the emergence of social value as a new currency in business transactions has been accompanied by considerable confusion as to what the term means. The debate around social value has become characterised by a multitude of different commentators attaching different definitions to the term from numerous perspectives including: government; the third sector; business; investors; and from many different fields in academia such as business, sociology, politics and accounting.

Some of the most widely cited definitions of social value are provided below to illustrate the potential confusion that can confront someone who is trying to come to terms with this emerging area. For example,

Emerson et al. (2000) defined social value as being created when resources, inputs, processes or policies are combined to generate improvements in the lives of individuals or society as a whole.

Cook and Monk (2012:11) define social value as 'the additional benefit to the community from a commissioning/procurement process over and above the direct purchasing of goods, services and outcomes'.

Social Enterprise UK (2012: 11) define social value in the context of social procurement as 'the additional benefit to the community from a commissioning/ procurement process over and above the direct purchasing of goods, services and outcomes'. According to Social Enterprise UK, social value involves looking beyond the price of goods and services procured to consider the collective benefit to a community when a public body chooses to award a contract.

The Social Value (Public Services) Act 2012 in the UK defines social value as a concept which seeks to maximise the additional benefit that can be created by procuring or commissioning goods and services, above and beyond the benefit of merely the goods and services themselves.

The UK Cabinet Office (2012) defines social value as the positive social, environmental and economic impact of an activity on stakeholders over and above what would have happened anyway, taking into account the negative impact of an activity.

Social Value International, a global network focused on social impact and social value, defines social value as the relative importance that people place on the changes they experience in their lives, from the perspective of those affected by an organisation's work, not all of which can be captured in market prices.

In the UK, Croydon Council's acclaimed Social Value Toolkit defines social value as arising from 'a process whereby organisations meet their needs for goods, services, works and utilities in a way that achieves value for money on a whole life basis in terms of generating benefits not only to the organisation, but also to society and the economy, whilst minimising damage to the environment' (Croydon 2012: 5).

Another prominent UK local council in the social value debate includes Knowsley Metropolitan Borough Council, which defines social value as 'Outcomes, measures and activity that will create strong and well-connected public, private and social sectors that enable communities to be more resilient'.

The Social Value Portal in the UK, an on-line tool that allows organisations to measure and manage the contribution that their organisation and supply chain makes to society, defines social value as the wider financial and non-financial impacts of programmes, organisations and interventions, including the wellbeing of individuals and communities, social capital and the environment.

The language of social value

The confusion around the concept of social value has been exacerbated by commentators using the term 'social value' interchangeability with other related terms such as 'social benefit', 'community benefit', 'social impact', 'social output', 'social outcomes' and the broader concept of CSR. The language of social value has become vague and imprecise with different terms meaning different things to different people depending on the context in which it is used. Indeed, in the UK, a review of the implementation of the Public Services (Social Value) Act 2012 by the UK Cabinet Office found that difficulties in defining social value are a major barrier to its implementation (Cabinet Office 2015). More recently, in the specific context of construction, Burke and King (2016) and Farag, McDermott and Huelin (2016) found that there was considerable confusion around what social value means in the industry due to a lack of guidance on how to deliver and define social value and a lack of prominence given to social value in public sector construction tenders.

While some argue that the lack of any formal, agreed or indeed legal definition of social value is not a problem because it allows for innovation in the delivery of social value to the community, others fear that if left too vague there is a danger that the concept will lack any practical meaning. So, to address the definitional problems which plague the field as noted above, and without being too prescriptive as to what forms of value can be included in a definition, we begin this chapter by defining some key concepts and terminology used in the social value literature. This will allow us to move forward in this book with a common understanding of the correct terminology and of what we mean by social value, whilst also recognising that all organisations create unique forms of social value, both good and bad, in their own unique ways.

These key terms which have emerged in the developing field of social value, can be found in the growing body of academic literature and the many international practical guidelines published on the subject of social value measurement such as Brouwers, Prins and Salverda(2010) Cabinet Office (2012), CSI (2014), G8 (2014), ICAEW (2015), NCVO (2013), NPC (2014), GECES (2014), NSW (2015). Although there remains no agreed international standard for social value measurement and some variability in the use of these terms across different disciplines, the following definitions are widely recognised in the emerging debate around social value and for this reason are adopted in this book.

> **Social 'Inputs'** = resources invested in activity/programme/intervention (financial, natural, intellectual, human, physical, social).
> **Social 'Activities'** = organisational activities and specific programme activities which are aimed at or/and have an impact on the lives of beneficiaries.

Social 'Outputs' = the direct and tangible products and services from an organisation's activities (e.g. number of people/hours trained, number/hours employed etc).

Social 'Outcomes' = the immediate, intermediate and long-term changes (both positive and negative) in people's lives as a result of an organisation's activities. These changes can be both 'primary' (in the target population) and 'secondary' (in the people delivering the activities and in the target population's families and wider communities) and can take many forms. These include changes to happiness, self-esteem, knowledge and skills, relationships, behaviours such as substance abuse and increased income, increased housing stability and increased security and safety.

Social 'Impact' = Social impact is the net outcome of the activities taking into account negative and positive effects, and various counterfactuals which include deadweight (what would have happened anyway), drop-off (reducing benefit over time), attribution (what else could have contributed to the change), displacement (what other benefits does the intervention displace/push aside), and substitution (losses for others who might have lost out).

This language and set of definitions align with the core values at the heart of our discussion on social value where we advocate a systems approach to tackling complex problems that recognises the need to take a holistic view of the many private, public, charitable, not-for-profit and community organisations and individuals involved in delivering social value to the community. The value of a systems approach to thinking about social value is that it provides a tool to understand and examine the linkages and interactions between the many components that comprise the entirety of the social value creation system. It brings together the design, construction and operation of that system as an integrated complex composition of many interconnected sub-systems, within the context of a wider community environment with specific social needs and priorities.

The economics of social value

Understanding of value in a business context has been traditionally informed by an economic perspective (see Griffith, Knight and King 2003; Kelly, Male and Graham 2007). Three widely used concepts include 'value in use', 'value in exchange', and 'esteem value'. Value in use refers to the function of a service or a product that satisfies a need or generates pleasure for its owner (Griffith, Knight and King 2003: 71). Value in exchange relates to the worth or the monetary sums for which the service or product can be traded (Kelly, Male and Graham 2007: 150). Importantly, a particular service or a product may have a great value in use but a relatively low value in exchange; for example, water. Alternatively, a service or a product may have a high value in exchange but a low value in use; for example, diamonds. This is known as the paradox of value (Griffith, Knight and King 2003: 71; Kelly, Male and Graham 2007: 152). It demonstrates that market prices often do not necessarily reflect personal notions of value – a problem which lies at the heart of controversies around the monetisation of social value using techniques such as social return on investment (SROI). In the context of the built environment, it is also helps to explain why market prices can often inadequately reflect the social value of construction and infrastructure developments where the value in use for various stakeholders is high but value in exchange is relatively low. A good example of this is social housing provision within a deprived area (for examples see the case studies in Chapter 9 in Part 2 of this book) where houses offer high social value by providing homes for occupants suffering disadvantage who may otherwise be homeless while financial market capitalisation opportunities for investors are small – a common problem facing many social entrepreneurs. Finally, esteem value relates to the functions of prestige, appearance, and/ or other non-quantifiable benefits, such as purchasing something simply for the sake of possession (Kelly, Male and Graham, 2007: 151). Architecture and design work often carry high esteem value. However, evaluation of such value is subjective, and thus very difficult to measure. Measuring value then becomes an exercise of appreciating and considering often complex and competing questions about value in use, value in exchange and esteem value. In practical terms, this raises questions about how we weight the value judgements of specific elements of a construction project for example for different stakeholders affected. In an attempt to address these types of dilemmas, concepts of public value and shared value have emerged as two specific ideas from value theory that promote socially aware value management (Moore 1995; Benington and Moore 2011).

At the heart of public value theory is a strategic triangle consisting of:

1. Clarity and specificity of what public value means: the strategic goals and public value outcomes which are aimed for in a given situation.
2. Authorisation: creation of an 'authorizing environment' necessary to achieving the desired public value outcomes, building and sustaining a coalition of stakeholders from the public, private and third sectors.
3. Harnessing and mobilizing the operational resources (finance, staff, skills, technology), both inside and outside the organisation, which are necessary to achieve desired public value outcomes (Benington and Moore 2011: 4).

Each of these three factors is strategically important, but they are rarely in alignment, and thus public managers have to strive constantly to bring them in to balance and to negotiate workable trade-offs between them (ibid). Much of the current thinking and policy debate around social value is aligned with the strategic triangle. For example, the Social Value Act in the UK is an exemplar of creating an authorising environment which requires public sector procurement efforts to articulate strategic goals and harness operational resources in ways that secure public value. Some local authorities in the UK actively embrace the idea beyond compliance with the Act, and define (public) social value as it applies to their specific contexts, mobilising operational resources to achieve those goals and outcomes.

The point of departure where discourse on social value differs from public value is rooted in its focus on a broader range of stakeholders and participants. Conceptually, public value is located within the public sector, and hence aligned with management of public sector projects and processes. Whilst social value is also often relevant to public sector work, it is equally relevant to private sector clients and contractors and the supply-chain, as well as the third sector (for example, see the case studies in Part 2 in this book). This brings us to the notion of shared value first articulated by Porter and Kramer (2006, 2011) to represent a new way of joined-up thinking about the relationship between economic, social and environmental goals. The concept of shared value questions whether trade-offs should be made between business and social activities and is based on the assumption of mutual dependence between business, sustainable environments and healthy communities. According to Porter and Kramer (2011) what is good for business is also good for society and verse versa. In other words, the relationship between social, environmental and economic goals is not hierarchical (where economic goals traditionally take first priority) but complementary. Porter and Kramer (2011) argue that all organisations can pursue shared value opportunities in three mains ways:

1. Re-designing an organisation's products and services to benefit communities and at the same time increase market share/profitability.
2. Making improvements to internal operations along the value chain which improves efficiency and also benefits society through reduced energy consumption, improved employee incomes etc.
3. Improving the external environment through strengthening local suppliers or subcontractors etc. which also improves productivity.

In both language and ideology, the concept of shared value shares much with the stakeholder management literature, for example stakeholder value and the Instrumental Stakeholder Theory (Jones 1995); blended value (Emerson 2000); and, social innovation (Kanter 1999). These concepts have made progress towards enhancing attention to the social dimensions of business, and may act as a spur for better practice (Crane et al 2014; Awale and Rowlinson 2014; Vogelius and Storgaard 2016). They begin with the assumption that social value is necessarily and explicitly a part of doing business, and so managers must articulate the shared sense of value their organisation creates to their many different stakeholders. It follows that organisations and their value base are located somewhere on a continuum that runs from those who are acutely aware of the social context within which they operate on the one hand, and those who do not take social value into consideration and measure performance strictly on financial and economic terms on the other hand.

Despite a growing number of businesses in the built environment embracing the shared value philosophy, critics argue that many companies are struggling to implement the shared value approach and that reports of success have so far relied on anecdote and feel-good stories rather than on empirical evidence (see for example Beschorner 2013, Williams and Hayes 2013). Critics also argue that the concept lacks a rigorous conceptual foundation and there is currently no widely accepted framework for measuring shared value in practice. Furthermore, it has been very difficult to gauge the success of the organisations that are said to champion this idea and critics argue that the concept of shared value will never maximise shared value because it places economic value generation on a par with social value creation. This means that social problems that do not have a business case will be at risk of being neglected.

Social procurement and social value

With all of the recent hype around the concept of social value, one would be for-given for thinking it is new. However, the concept has existed for many decades in business, promoted in large part by the long-standing use of social procure-ment policies and legislation in Europe, the UK and the US to promote fair labour practices and ethnic and other minority group inclusion and employment (McCrudden 2004). It is notable that most of these policies target the built envir-onment as a major force for change because of the large public procurement spend in this sector and the relatively large multiplier effect of the built environment into the wider community (Loosemore and Higgon 2015). For example, in the US, The Public Law 95–507 Act of 1978 has long required firms tendering for con-struction contracts of over a certain value to submit a buying plan that includes percentage goals for employing minority businesses. However, as Barraket, Keast and Furneaux (2016) note, there has been renewed focus on the concept of social value in recent years, as interest in the potential value of social procurement has grown as an innovative mechanism to create wider economic, social, cultural and environmental benefits for society beyond the normal purchasing of good and services in the public sector. According the Barraket, Keast and Furneaux (2016) this has been driven by a number of major trends including the emergence of what they describe as the 'audit society' and by recent trends in 'new public govern-ance' with its outcomes-based procurement, relational contracting, public private partnerships, outsourcing and networked forms of governance which rely on cross sector collaboration between government, business, third sector and community organisations. The growing global interest in the positive role that cross-sector collaboration between the private, public and the third sectors can play in building more sustainable societies has been operationalised by emerging social procure-ment policies, legislation and regulations, to which the construction industry must now respond when tendering for public infrastructure and construction projects.

One prominent example of the above trend is the Public Services (Social Value) Act (2012) in the UK, which was designed to transform the way that public bodies in England and Wales buy products and services through a broadened definition of 'value for money' which goes beyond price to incorporate wider social, cul-tural, economic and environmental benefits to society. Although these broader concepts of 'value for money' were first promoted within the 'best value' regime, introduced under the Local Government Act 1999 in the UK, the Social Value Act 2012 has been important in introducing a formal 'duty' on public bodies to consider the economic, social and environmental well-being of a relevant community in conducting the process of procurement, and was a catalyst to social value becoming a formally recognised term in the UK government procurement context.

Social value has also been formalised in the European Union (EU) by various EU directives on the coordination of procedures for the award of public works contracts such as Directive 2004/17/EC and Directive2004/18/EC, which play a key role in the Europe 2020 Strategy. The EU has issued guidance to its member states around socially responsible public procurement which proposes that European Governments should encourage smaller social enterprises to tender for work which is normally dominated by large business (for example through unbundling contracts into smaller packages) (EC 2010). The EU's focus on social value has also been highlighted in its 2011 publication 'Buying Social – A Guide to Taking Account of Social Considerations in Public Procurement', which argued that having created a common market, it was time to make better use of public procurement in support of common societal goals by promoting it in EU supply chains. This means employment opportunities for disadvantaged groups; decent working conditions; compliance with human rights and social and labour laws; social inclusion; opportunities for third sector organisations; ethical trade; corporate social responsibility; stimulating socially conscious markets; and socially responsive governance.

In contrast to the deliberatively flexible and non-prescriptive Social Value Act in the UK, other countries like Australia, Canada and South Africa have been more specific in their social procurement requirements by targeting specific disadvantaged groups, at both State and Federal level. For example, the Australian Commonwealth Government's Indigenous Procurement Policy (Commonwealth of Australia 2015b) has set a target that three percent of new domestic Commonwealth Government contracts will be awarded to indigenous suppliers in 2019–2020, resulting in more than 1,500 contracts by 2019–2020. While indigenous businesses can bid for any contracts, departments must look first to indigenous businesses for all building, construction and maintenance contracts in remote areas, regardless of value, in addition to all domestic contracts valued between $80,000 and $200,000. Mandatory minimum requirements also apply to all building, construction and maintenance contracts valued at $7.5 million or more, which include a contract-based requirement to achieve at least 4 percent indigenous employment and/or supplier use on average over the term of the contract; or an organisation-based requirement to achieve 3 percent indigenous employment and/or supplier use across the organisation on average over the term of the contract. In addition, where part of the contract is to be delivered in a 'remote area', the government agency and the contractor will agree to significant indigenous employment or supplier use requirements in that area. Any contractor tendering for a commonwealth government agency building, construction and maintenance project must describe how they will meet these targets and report on them regularly at ultimate risk of sanctions for under-performance. Similarly, South Africa's Preferential Procurement Policy Framework Act (2000) and Preferential Procurement Regulations (2017) have set a target of at least 30 percent of the supply chain to small and micro businesses which are majority 51 percent owned by black people of various disadvantaged backgrounds such as youth, women, rural areas, and people with disabilities. Since 1996, the Canadian Government has also operated a Procurement Strategy for Aboriginal Business (PSAB) 2016, which aims to increase the number of aboriginal firms participating in federal government procurement processes. The PSAB applies to all federal government departments and agencies and like Australia and South Africa includes various mechanisms such as 'set asides', where a percentage of opportunities are reserved for a disadvantaged or minority group. The PSAB is open to all aboriginal businesses that are at least 51 percent owned and controlled by aboriginal people; has six or more full-time staff; and employs at least 30 percent aboriginal employees.

Sitting underneath the government policies discussed above, most countries also have additional layers of social procurement policies developed by progressive state governments, local councils and individual clients and associations representing specific clients groups at a more local level. Notable examples include the Croydon Council Social Value Toolkit in London UK (Croydon 2012), Toronto's Social Procurement Framework in Canada (2014), the Social Procurement in New South Wales Guidelines in Australia (2012) and the Social Value Procurement Toolkit produced by the Housing Associations Charitable Trust in the UK (HACT 2016). These layers of developing social procurement laws, policies and guidelines have a strong focus on construction and infrastructure spending and represent a constantly shifting and complex social value creation landscape with which construction professionals need to comply.

Most recently, to provide a language by which people can talk about social procurement and social value, we have also seen the development of international soft instruments such as ISO 20400 *Sustainable Procurement – Guidance* the world's first International Standard for sustainable procurement which aims to help organisations develop and implement sustainable purchasing practices and policies. Sustainable procurement entails making purchasing decisions which take into account social, economic and environmental factors, in a way that collectively benefits an organisation, society and the environment. This requires that firms in supply chains behave fairly, responsibly and ethically and that the products and services purchased are sustainable. ISO 20400 provides guidelines for integrating sustainability into an organisation's procurement policy, strategy and process, defining the principles of sustainable procurement such as accountability, transparency, respect for human rights and ethical behaviour, although it refers back to ISO 26000 International Guidance on Social Responsibility (2010) for detailed examples and indicators in these areas. So, for example, it may not be acceptable for a construction company to claim a project is creating jobs for people in a local community if it involves destroying the natural environment to do so. This was recently vividly demonstrated in Sydney's highly controversial new light rail project, which will provide numerous economic benefits to the Sydney community and potentially cut road pollution, but which also involved the destruction of social, cultural and environmental heritage and assets associated with the felling of numerous 100-year-old fig trees, which resulted in vociferous community protests (Aubusson 2016). Finally, although this is a continually developing area, there are numerous international standards such as The Global Reporting Initiative (GRI), Integrated Reporting Framework, United Nations Sustainable Development Goals and Sustainable Stock Exchanges Initiative to help organisations monitor and report performance, strategic and governance information clearly, concisely and in a comparable format which reflects the commercial, social and environmental context in which they operate. Whilst these types of high-level instruments affect only the largest firms in the construction industry, and can be legally ignored by those not listed on the stock exchange, it is an extremely unwise business that does so because many major clients of construction who do need to comply, are increasingly requiring their construction supply chains to do the same. The trickle-down effect of all of the above policies into the construction industry supply chain is therefore inevitable and will eventually impact everyone working in the construction industry no matter how large or small. Indeed, recent research in Australia by the Australian Council of Superannuation Investors (2017) shows that the number of listed companies engaged in sustainability reporting has almost tripled since 2008, with real estate companies being listed as one of the leaders across all sectors. This is a strong sign that construction clients will increasingly demand those in their supply chains to demonstrate they are contributing social value to the communities in which they will build their buildings.

Loosemore and Lim's (2017) recent review of CSR literature in construction suggests that there are many potential commercial benefits for firms that take a leadership role in this emerging area, although these benefits are not as widely perceived in the built environment as they seem to be in other industries. These include:

- competitive advantage (with socially responsible clients);
- demonstrable corporate citizenship and social responsibility;
- improved employee recruitment, engagement and retention (people want more out of work than work: new knowledge, motivation, satisfaction, self-worth, self-confidence, new skills like mentoring etc.);
- compliance with growing social procurement requirements;
- improved community engagement and public relations;
- greater innovation in bids (new leverage, more diverse supply chains, more supplier options, new networks and new innovative ideas brought by social enterprises);
- widening markets and customer-base;
- positive reputation (communities, clients, shareholders, employees and other stakeholders);

- building corporate loyalty and positive brand awareness with both internal and external stakeholders;
- investment – increasingly financial investment decisions are made on non-financial performance.

In maximising these benefits, the growing numbers of social procurement frameworks, guidelines and toolkits emphasise the importance of:

- starting from the needs of the community using a bottom-up approach;
- understanding the community and its disadvantaged cohorts you are seeking to help – their needs, aspirations and challenges;
- keeping social procurement simple, transparent and fair to all suppliers – both new and existing;
- understanding how to navigate across government, business, third and community sectors and how to build effective cross-sector collaborative alliances to deliver community needs;
- understanding and managing the significant risks and opportunities of engaging with the social sector and the beneficiaries you seek to help;
- monitoring, measuring and enforcing social value outcomes through rewards and sanctions.

(See for example Nicholls et al 2005; Bonwick and Daniels 2014; Burkett 2010; Croydon 2012; Newman and Burkett 2012; Furneaux and Barraket 2014; Barraket, Keast and Furneaux 2016; LePage 2014; HACT 2016.) These practical issues are discussed in more detail in subsequent chapters.

The political context of social value

While there are many passionate advocates of the social value debate, it is critical to understand the political context in which this book is set. First, delivering social value to a community will inevitably involve working with a wide variety of stakeholders across a range of sectors (private, public, community, not-for-profit, charity, etc.) with a wide variety of interests, priorities and relationships, which will need to be carefully navigated to identify and deliver the outcomes the community needs. Most construction firms are simply not equipped or experienced enough to be able to do this effectively and will require specialist advice from people who have worked across these sectors to do so. Such people are not easy to come by.

Second, the idea of involving the private sector in tackling social problems and ultimately welfare provision is inherently political and highly controversial in itself. For many critics, the process of involving private firms in the delivery of welfare to society, under the banner of social innovation, social enterprise and social procurement, is simply a rhetorical smoke-screen for the dismantling of the welfare state and justifying government austerity programmes which have made deep cuts to welfare (Macmillan 2013; Doherty, Haugh and Lyon 2014; Whelan

2012). Indeed, Grisolia and Ferragina (2015) describe social innovation as yet another politically convenient buzzword which inserts the term 'social' into the business lexicon to avoid heated discussions about the social impacts of austerity and growing structural inequalities in society, under the guise of neoliberal orthodoxy – liberalisation, deregulation, devolution, individual or group empowerment. Barraket and Weissman (2009: 4) also warn that when poorly designed and implemented, social procurement policies can lead to an 'inefficient mix of production across the economy' and Esteves and Barclay (2011) argue that they can also lead to unintended adverse social impacts and encourage perverse market behaviour. For example, in the Australian construction industry, Denny-Smith and Loosemore (2017) warn of unscrupulous businesses setting up scam indigenous companies to access significant government social procurement budgets dedicated to this area. Esteves and Barclay (2011) point out that community resentment and dissatisfaction can also result from tokenistic compliance with social procurement policies when firms motivated by compliance rather than a genuine concern for disadvantage provide only menial jobs to target groups in order to meet social procurement imperatives. Ironically, third sector organisations, which become part of the construction sector's diversified supply chains, can also become too dependent on what is a temporary opportunity and be left vulnerable to the business cycle when a social procurement initiative ends. It must also be remembered that other local businesses, which may be dislodged from supply chains by preference being given to third sector organisations, also provide important employment and social benefits for society, and may be more efficient in doing so. Finally, perhaps most concerning is the current lack of empirical evidence that third sector organisations are any better at delivering social value than normal construction subcontractors. The answer to this critical question is made more complicated by a lack of agreement and discipline around the practice of measuring and communicating social value, which ensures that costs and benefits of engaging in social procurement cannot be reliably established (Burkett, 2010; LePage, 2014). We discuss this more in Chapter 5. However, it is this lack of methodology and evidence that has provided fertile ground for critics to argue that there is little empirical evidence to support (or deny) the many claims that these relatively new approaches to creating social value offer new advantages to society over traditional government welfare systems.

The opportunity for industry leadership in social value

Despite the inevitable politics and the many barriers to social value creation in the highly commercial built environment, the timing for someone to 'step-up' and show some leadership in this area is perfect. Built environment can do more than most to address growing inequity and disadvantage in society; companies working in the construction industry, in both a consulting and contracting capacity, can benefit more than most. Not only is there a growing social need, as evidenced by growing inequity and disadvantage in many societies (Jenkins 2015; Commonwealth of Australia 2015, 2017; OECD 2017), but as noted above, governments, communities and socially responsible private clients are increasingly expecting construction firms to meet their corporate citizenship responsibilities. There is also an unprecedented global building and infrastructure pipeline to leverage and a deepening skills shortage that can be addressed using social procurement to access a more diverse workforce including women, people with disabilities and ethnic groups who have been traditionally excluded from the built environment due to ingrained stereotypes about the qualities of the perfect construction employee. Indeed, the global construction sector employs more people than any other industry and is anticipated to grow by more than 70 percent to $15 trillion worldwide by 2025 (WMI 2010; GCP 2013). Estimates suggest that 50 percent of all construction occupations in Australia will be in shortage over the next 5 years (CICA 2015). To meet emerging construction and infrastructure spending, and to replace an ageing workforce, the Australian construction industry is said to need an extra 13,000 to 15,000 new apprentices per year and an additional 300,000 skilled workers nationally over the next decade; a 30 percent increase on the current workforce of 1,033,000 people (Master Builders Association 2017; ABS 2017). In the UK, serious skills shortages are also predicted and are likely to be exacerbated by Brexit. Indeed, Farmer (2016) identifies skills shortages as the biggest constraint on UK industry meeting urgent UK housing needs of 250,000 homes per year. Unlike many other industries, the construction industry also operates in some of the world's most disadvantaged areas and its multiplier effect into the wider economy is one of the largest of any industries. For example, the Australian Bureau of Statistics (ABS 2002) estimates that for every $1m spent on construction output a possible $2.9m in output would be generated in the economy as a whole, giving rise to 9 jobs in the construction industry (the initial employment effect) and 37 jobs in the wider economy. In the UK, the construction sector's economic multiplier, at around £2.84, is one of the highest of any sector as a result of the industry's relatively low level of imports, its extended and complex supply chains and its relatively high labour intensity (RICS 2011). There is also a desperate need for the built environment in many parts of the world to improve its tarnished reputation, which Transparency International (2011) and Ernst and Young (2012) describe as having the largest global propensity for unethical behaviour. Finally, the built environment offers many untapped opportunities for disadvantaged groups because unlike many other industries it operates in many remote and disadvantaged communities and its employment structure of skilled and unskilled trades, often dominated by certain cultural groups (Italian concreters, Croatian carpenters, Korean tilers and Chinese plaster boarders, etc.), offers support structures for numerous disadvantaged groups such as refugees (Loosemore et al. 2001). In Australia, the construction industry is the largest youth employer with 43.3 percent of workers aged 15 to 34 years, compared with 38.8 percent across all industries (CICA 2015) including a disproportionately large number of indigenous businesses operating (yourtown 2017).

Defining the social value proposition of a construction project

In the context of the above discussions, our definition of social value is simply the 'social impact' any construction organisation, project or programme makes to the lives of internal and external stakeholders affected by its activities, including those working in the industry and in the communities in which it operates.

One of the unique features of the construction industry, compared to other industries such as manufacturing for example, is that its work is undertaken through projects with high levels of variability and change in design, location, production environment and supply chain structure. This ensures that every construction project offers a different social value proposition, depending on a range of factors such as:

- Project size – larger projects typically present the opportunity to offer more social value than smaller projects.
- Project type – economic infrastructure such as roads and rail offer enormous social value opportunities because of the many communities affected. Social infrastructure (schools, hospitals, etc.) offer social value by the very nature of their operations, for example in education and health. Housing offers other forms of social value associated with security, comfort, a sense of home, family cohesion and wealth creation.
- Project location – projects in areas of low socio-economic status typically offer a higher social value proposition because of greater community needs and the opportunity to employ locally disadvantaged people.
- Project design – innovative urban and building design encourages safe and healthy living and generates economic value by attracting new business to an area. The materials specified in design can be sourced from responsible suppliers which create social value in their production and the construction technologies implied through design can create or destroy employment opportunities.
- Procurement approach – integrated procurement reduces the split-incentives that create short-term thinking and offers more in-built incentives for sustainable solutions (environmentally sustainable technologies such as energy efficiency reduce life-cycle costs). Integrated procurement also offers more opportunities for stakeholders (such as building users and communities) to share and collaborate in the project outcome.
- Local supply chains – the use of local businesses and social enterprises in the project supply chain can create employment opportunities and business for local and disadvantaged groups.
- Timing (the political, social and economic environment in which a project is built) – projects built during recessions can create more social value by keeping people in work and if the government is politically open to account for social value in its tender processes then there is more incentive and opportunity to create social value in the project.

- The client – socially responsible clients offer more incentives and opportunities for the supply chain to create social value.

Understanding the above factors is important for managers in prioritising limited resources to achieve maximum social impact, as is limiting the types of social value impact they wish to create. Loosemore and Phua (2011) suggest that to maximise the social impact of construction activities, it is important to not try to be everything to everyone and instead to focus one's resources on:

- only a few strategic areas (working within existing resource and time constraints and community social needs and priorities);
- areas of impact that are of concern to primary stakeholders;
- areas of impact that align with organisational mission, values and core business goals;
- areas of impact that are sustainable and that can be supported and maintained in the long-term;
- areas of impact where there is previous experience and advantage over other providers through access to inimitable core resources.

However, the almost infinite range of social impacts potentially created by construction projects can often be overwhelming and in order to structure thinking, it is useful to refer to social impact frameworks such as the abovementioned ISO 26000 International Guidance on Social Responsibility. The value of this framework is that it was developed across multiple stakeholder groups (private firms, public organisations, NGOs, academic researchers etc.) and lists seven core CSR categories which can be used to categorise, focus resources on, and ultimately measure the social value which can be created by construction projects. Kritkausky and Schmidt (2011) describe the types of activities that are typically included in these categories as follows:

1. Community involvement and development – includes social value created from activities such as: community engagement; supporting local charities and causes; being a good neighbour and; providing training, employment and business opportunities for community members and local businesses through local purchasing and employment.
2. Human rights – includes social value created from activities such as: establishing fair mechanisms for promoting human rights; equity and diversity policies; respecting individuals' rights to freedom of association, opinion and expression; and respecting economic, social and cultural rights.
3. Labour practices – includes social value created from activities such as: providing a just, safe and healthy work environment for employees which involves: good wages and working conditions such as pensions, holidays, work-life balance, sick pay and social protection; support dialogue between employers and employees; and providing opportunities for human resource development.

4. Environmental – includes social value created from activities such as: pollution prevention; emissions reduction; use of sustainable renewable resources; life-cycle management; using environmentally sound technologies and practices; and sustainable procurement.
5. Fair operating practices – includes social value created from activities such as: respecting the law; practicing accountability and fairness in business relationships; social procurement; and responsible sourcing.
6. Consumer issues – includes social value created from activities such as: providing healthy and safe products; giving accurate information about products and services; promoting sustainable consumption; designing products which can be reused, repaired or recycled; reducing packaging waste; and protecting consumer privacy when handling personal data.
7. Organisational governance – includes social value created from activities such as: accountability and transparency in decision making; respecting laws; responsible use of financial, natural and human resources; considering all key stakeholders in decision making including minority groups; monitoring and reporting of business' activities, both positive and negative.

The social value created in each of the areas listed by ISO 26000 can take a multitude of forms for a wide variety of stakeholders depending on the scale, scope and reach of the programmes, projects and organisational activities involved. These can include and are not limited to:

- increased training and employment opportunities and income for disadvantaged groups;
- increased knowledge, skills and qualifications through training and education;
- increased equality of opportunity, for marginalised and disadvantaged groups;
- increased community resilience, cohesion and security;
- reduced demand for public services;
- increased income and productivity through employment and entrepreneurship;
- increased tax revenues for government;
- reduced poverty;
- increased hope, self-respect and confidence for the future;
- increased integration, contribution and belonging to normal society;
- improved mental and physical health (reduced health costs);
- improved income and wealth (reduced dependence on social security and welfare support services);
- improved social behaviour such as reduced crime and re-offending (reduced incarceration and recidivism costs), reduced substance abuse, suicide and family violence (reduced policing and intervention costs) and reduced anti-social behaviour (reduced costs associated with issues like graffiti and environmental degradation);

- spill-over multiplier benefits into families leading to improve community resilience (more harmonious community relations and more money into communities to stimulate local/regional/remote economies).

Creating social value through the construction project life-cycle

One of the limitations of the current debate on social value in the built environment is that it is too heavily focused on the construction stage of projects, because this is where new emerging social procurement policy is being aimed. However, it is important to understand that social value is created over the entire life-cycle of a construction project including the early urban planning, project planning, design and operational phases. Indeed, the earlier in a project one starts thinking about social value the better. There is already a significant body of literature in the fields of urban planning and design, heritage and architecture, which addresses the potential social value creating opportunities in design and planning, even though it might not specifically use the term social value to encapsulate these ideas.

Creating social value in the design stage

There is indisputable evidence to show that well designed urban places, buildings and infrastructure can both transform the social fabric of local communities and enhance economies at regional and national level. The 2012 Olympics' impact on London's East End in the UK, and the Guggenheim Museum's role in revitalising Bilbao in northern Spain are excellent examples of how well-designed building and infrastructure projects can regenerate disadvantaged communities by creating a new sense of identity, place and pride. As Johnston (1992) argues, the social value we attach to our surroundings and built environment are related to more than just their physical form. Buildings and places are valuable environments where we meet, enjoy, relax, commemorate, work and trade and they are also spaces to which we attach strong and often shared emotional, spiritual, nostalgic, political, cultural, historic, social, aesthetic and economic meanings which embody our identity, ideas and ideals (our sense of place). Recognising this, good designers have long embraced the concept of 'placemaking'. This collaborative, multidisciplinary and community-driven approach to the planning and design of urban environments, which is developed further in Chapter 6, harnesses a community's knowledge, ideas and resources as a way of creating urban environments that promote people's health, wealth, happiness, and wellbeing (Fleming 2007). More recent research in architecture, urban design and planning is providing compelling evidence that good building, infrastructure and urban design provides a wide range of economic, social, cultural, environmental, security and health benefits to the community by providing:

- affordable housing for people to live in;
- efficient and accessible infrastructure for people to travel on;
- commercial space for businesses to thrive and provide employment;

- safe places which minimise opportunities for crime;
- environments which promote health and reduce noise, emissions and pollution by encouraging walking rather than car use; and by
- spaces for people to meet, socialise and relax (see for example, Barton, Thompson and Grant 2015; Anderson et al. 2016).

As research by the Commission for Architecture and the Built Environment (CABE) (2002) shows, people attach great value to the built environment in which they live. An overwhelming 81 percent of people surveyed indicated that they were 'interested in how the built environment looks and feels', and over 30 percent said that they wanted more of a say in the design of buildings and public spaces around them. A high proportion (85 percent) of the people surveyed also agreed that better quality buildings and public spaces improve the quality of people's lives and that the quality of the built environment made a difference to the way they felt. Every type of building and infrastructure seems to have the potential to create social value. For example, CABE (2002) showed that people work more productively in well-designed commercial and industrial buildings. This important benefit stems from the way design affects numerous factors: occupancy levels, employee satisfaction, mental and physical health and wellbeing and teamwork, time required for handling and transporting products and rental rates for commercial offices. CABE (2002) also showed that well-designed schools improve education results by affecting staff morale and pupil motivation through creating spaces that facilitate effective learning and educational outcomes and even the level of supervision needed during breaks. This, in turn, was found to lead to improved pupil behaviour and more time for staff to engage in learning activities. In a hospital context, CABE (2002) showed how the design of hospitals can make a difference to how rapidly patients recover, how many drugs they need, to the rate of infections and post-surgical complications, patient and staff satisfaction and even levels of violence against staff. Finally, well-designed houses produce many positive social outcomes for their occupants linked to energy efficiency, maintenance costs, accidents and health, security and crime, safety and fire risks, family cohesion, sense of pride and property values.

Good urban design can also create economic value. According to research focused on industry innovation clusters such as Silicon Valley, good urban design can increase levels of productivity and innovation in the economy by creating communities of practice which allow people to share information across disparate industry boundaries, exposing people to divergent ways of thinking and enabling ideas from one discipline to be imported into others (see Horrigan 2011; Mitchell 2012). These authors provide many examples of how cities have shaped the development of ideas and stimulated innovation and economic prosperity over the centuries. For example, Florence, Italy, is known to have created the environment for the start of the Renaissance in Europe by providing public spaces where likeminded artists, scholars and philanthropists could meet to discuss new ideas during the late fourteenth century. The Australian gold rush in the late nineteenth century drew in thousands of Chinese migrants who eventually established the Chinatowns of Melbourne and Sydney, which have since acted as incubators of Chinese entre-preneurship. Similarly, in the US, the island of Manhattan became the home of many European migrants after the Second World War, creating a breeding ground for many of America's most powerful media and advertising entrepreneurs.

However, it is not just the design process itself and the resulting designs that can create social value. Materials specified in design can be sourced from responsible suppliers, which create social value in their production. While the environmental impacts of materials specified by designers have long been scrutinised, more recent research and activism has focused on the social impacts of design decisions. For example, recent reports show that modern slavery in construction supply chains is a serious problem as a result of materials being sourced from countries with dubious human rights records that abuse workers' rights, employ child labour and are guilty of practicing modern slavery (Supply Chain Sustainability School 2018). Furthermore, the construction technologies embedded in the design can create or destroy employment opportunities. For example, designs that use prefabricated components can shift jobs off-site into factories, reducing waste, improving safety and opening opportunities for the industry to employ non-traditional groups nor-mally excluded from the construction workforce such as people with disabilities. On the other hand, while the use of off-site prefabrication can create new job oppor-tunities, it can also destroy jobs in traditional construction trades and occupations. Recent research in social procurement in the construction industry suggests that whilst construction contractors are currently the focus of government social pro-curement policies, consultants such as designers, engineers and planners will soon be subject to the same scrutiny (Loosemore and Higgon 2015).

Creating social value in the construction stage

There are many ways in which those involved in the construction stage of projects can contribute social value to the communities in which they build. The Considerate Constructors Scheme in the UK (a non-profit, independent organisa-tion established to improve the construction industry's image through a voluntary code of practice) provides numerous examples of how this is being done in prac-tice. Whilst there are too many excellent initiatives to include here, one useful example demonstrates how a major UK building contractor has helped homeless people sleeping adjacent to its site join the workforce by providing work experi-ence in partnership with a local homelessness charity. The UK Cabinet Office's Review of the UK's Social Value Act (Cabinet Office 2015) also documents many examples including a UK Council that unbundled its construction contracts into smaller packages to enable local businesses to bid for its projects. Balancing the extra costs associated with this strategy against the extra social value created, the Council also opted not to apply a financial turnover threshold or credit score, which had traditionally represented another barrier to small businesses prequalifying to tender for construction work.

Europe's largest infrastructure project, Crossrail in the UK, is also seen an exemplar of social value creation in practice. Although much of the social impact from this project has never been accurately measured, it was innovative in many ways. For example, the client required each Tier-1 contractor to create a Responsible Procurement Plan (with targets proportionate to the size and nature of individual construction packages) and to meet and report quarterly on a number of social value creation objectives, including:

- labour needs and training targets (job starts, apprenticeships, work placements etc.);
- payment of the London living wage; use of the Crossrail jobs brokerage service to advertise all vacancies (including in its supply chain);
- equality and diversity in its workforce and supply chain including the use of small to medium sized enterprises;
- workforce welfare;
- understanding and positive management of social and environmental community impacts and;
- ethical sourcing of construction products and services.

Although there has been no independent research or data collected into the real 'social impact' of Crossrail's initiatives, taking into account counterfactuals such as what would have happened anyway, and recognising that up-to-date data on social outcomes is not publicly available as yet, the Crossrail Sustainability Report 2016 claims that 96 percent of Crossrail contracts were awarded to UK companies, 62 percent of Tier-1 contracts have been awarded to SMEs, 4,544 job starts had been created for local and previously unemployed people, 1,109 young people had undertaken work experience, drivers has attended 662 driver training courses and over 15,000 people had been trained at Crossrail's dedicated £13m Tunnelling and Underground Construction Academy. By April 2017 the project had created around 55,000 full-time equivalent jobs with a further 699 apprentices employed by Crossrail's train manufacturer, Bombardier and train operating company MTR. In addition to a 'target zero' ethos in workforce health and safety, and numerous environmental initiatives, other initiatives which created social value include community art programmes, corporate volunteering and workforce diversity programmes and a learning legacy programme that documents lessons learnt in areas such as project management, health and safety, environment, information management, engineering and sustainability (including social sustainability). This legacy programme is an important potential source of residual ongoing social impact, allowing lessons to be transferred to other major infrastructure projects including Paris Metro, Hinkley Point C, High Speed 2, Tideway and Heathrow third runway. Lessons documented by Blin and Eldred (2016, 2017) in this project legacy material include the need to:

- Separate social sustainability from traditional human resource management functions.
- Create a collaborative forum for Tier 1 contractors and the client to collectively set goals and systems to deliver social value.

- Establish clear up-front strategic social value priories from the start of a project, to make these contractual requirements and link them to payments and sanctions around performance.
- Enforce a strong, practical and flexible performance measurement, monitoring and assurance framework (based on both qualitative and quantitative measures) linked to specific and measurable contract requirements that supports and encourages performance beyond minimum requirements.
- Highlight, reward and share good practice.
- Give successful Tier-1 contractors enough time after contract award to fully develop their Responsible Procurement Plans (and to keep updating them through the contract).
- Ensure necessary systems, policies, procedures and resources are in place to deliver the stated goals.
- Make sure that there are requirements for the entire supply chain in this process; not just Tier-1 contractors.
- As a client, intervene more directly in supporting the delivery of social value targets, through initiatives such as jobs brokerage services and supply chain capacity building, rather than just leaving it to the supply chain.
- Work collaboratively with other clients in the industry to promote social sustainability.

In Australia, Multiplex's Connectivity Centres are an innovative and rare example of successful cross-sector collaboration during construction. The case study in Chapter 8 documents these Connectivity Centres in more detail and explains how these centres bring together government, community, not-for-profit, private businesses and charities, to collaboratively create social value in the communities in which Multiplex builds. Based on the premise that social problems are too complex for one organisation to solve alone, different cross-sector configurations, relationship, collaborations, joint ventures and partnerships are forged in different project locations in response to local community needs. These needs represent the starting point for the formation of each individual Connectivity Centre which, while having a physical presence on site where all these parties can meet and produce social solutions, is essentially defined by a collection of cross-sector relationships formed over many years of close collaboration, configured in a way which are matched to the needs of the community in which the project is being built.

Notable social value creation initiatives are not confined to large multinational contractors like Multiplex and mega projects like Crossrail. Small construction projects, builders and subcontractors can also create significant social value in the communities in which they operate. We showcase the work of two such organisations in Chapter 9. Indeed, as Loosemore and Phua (2011) note, in many ways these organisations are more closely embedded in their communities and see social value as just a normal way of doing business. For example, in the UK, SOAR Build Ltd is a small social enterprise which delivers construction trades services such as wall insulation, plastering, tiling, painting and general contracting with the primary aim of providing employment, education and on-the-job skills training opportunities to local long-term unemployed people from a range of disadvantaged backgrounds including ethnic minorities, women, ex-offenders and people with disabilities. Similarly, in Australia, a youth charity called yourtown operates a range of small social enterprises that deliver general building services and grounds and building maintenance services to public and private clients. yourtown are one of the very few construction organisations in the world to have taken the trouble to precisely measure and report the social impact their construction social enterprises create. For example, Bartlett et al.'s (2012) social impact analysis of yourtown's programmes showed that by giving disadvantaged youth more jobs, considerable social value is created for many stakeholders in the community, including: young people (improved self-confidence, self-esteem, hope, income, housing security, mental and physical health, drug and alcohol addiction); government (reduced crime rates and incarceration costs); and the community (improved sense of cohesion and inclusiveness as young people stay in their communities rather than move away).

Creating social value through the operations and facilities management stage

Traditionally the operational and facilities management phases of projects have featured prominently in documented stories of social value creation in the built environment. This is due to the availability of numerous work opportunities in areas like small building works, security, cleaning, grounds maintenance and landscaping, and concierge services, where Third Sector Organisations have traditionally operated. For example, in Australia Transfield Services (a global operations, maintenance and construction services business) has a long history of delivering social value through its facilities management contracts on large infrastructure projects in regional communities. Transfield Services use a range of social procurement strategies which engage numerous social enterprises in the provision of grounds, hospitality and catering services for indigenous people, the long term unemployed and those with disabilities who are often excluded from the jobs market. In the UK, Landmarc have also been a leader in the field of social value creation, working closely with Social Enterprises UK to actively measure and document the social impact of its activities. In its first social impact report (Masom et al. 2013) Landmarc reported that 51 percent of its supply chain expenditure was with small-to-medium sized enterprises and that social enterprises represented 60–70 percent of the organisations it contracted with. Landmarc also encourages entrepreneurial capacity in its local communities through its Rural Enterprise Hub and engages in cross-sector partnerships with organisations such as Recovery Careers Services to support wounded, injured and sick ex-service officers into employment. It's Rural Apprenticeships Scheme also supported more than 20 apprentices in its first year of operation and its £100,000 kick-starter fund called Landmarc 100 has provided financial support and mentoring for up to 100 start-up rural enterprises. In Australia, Burkett (2010) documents the Victorian Department of Human Services as an excellent example of using 'place-based' social procurement to ensure that the tenants on its most socially disadvantaged housing estates are provided with jobs and training through the use of social clauses in its housing estate maintenance contracts, creating over 650 jobs and 1,300 training opportunities in a five-year period. The Victorian Department of Human Services also developed a public tenant employment programme which was a joint venture with a major charity (Brotherhood of St Laurence) to deliver concierge services to seven high rise housing estates in Melbourne, thereby increasing employment and training opportunities for tenants and improving their physical environment, sense of community and safety. In the UK, Knightstone Housing Association also uses social clauses in its maintenance and repairs contracts to help its social housing tenants access work opportunities during the operational phases of its projects. Bidders are required to commit to certain percentage targets to help social housing tenants access training and work opportunities and to report on several outcomes including number of job opportunities, number of people into work, availability of work experience/volunteering placements, provision of courses and training, number of community activities, and the amount of money paid for Knightstone to deliver community activities on their behalf (Cabinet Office 2015).

The ethics of social value and what it means for measuring social value

While the above sections provide many examples of social value and the various forms it can take, to fully understand the concept of social value, one must appreciate that it is socially constructed and context dependent. In other words, different people involved in the creation of social value will see it and define it in many different ways. Furthermore, this will change over time in response to a multitude of situational factors such as personal circumstances and the social, political, economic and cultural environment in which a judgement about value is being made.

Theories of how people attribute 'value' to other people, actions, objects and situations go back to the very origins of philosophy and in particular the field of ethics and the work of ancient philosophers such as Plato and Aristotle, German philosopher Immanuel Kant and more recently American philosophers such as Ralph Barton Perry (Perry 1967; Barnes 1995; Schneewind et al. 2002; Anderson 1993; Wiggins 1987). At the heart of ethics are our beliefs about what is right or wrong and good or bad and a concern about someone other than ourselves. These beliefs, which lie at the heart of how we attribute value to something, develop over many generations in our communities, families, religions, philosophies, cultures and institutions, are continually changing and are instilled within us all through socialisation, education and our membership of various organisations, community groups and societies. The field of ethics offers us ethical rules, moral maps and principles that we can then use to find our way through difficult issues and provides managers with frameworks for making and evaluating business decisions, particularly in ambiguous situations where the interests of different stakeholders, moral values and organisational rules might appear to clash (Brown 2005). Table 1.1 lists different schools of thought that have evolved in the complex field of ethics and shows how just how easy it is for decision-makers to construct an ethical argument for-or-against any new idea and how ethical codes can clash in making value

Table 1.1 Common ethical schools of thought

School of thought	Basic principles
Supernaturalism	Supernaturalism draws its principles from religion and teaches that judgements about value are purely determined by religious doctrines.
Intuitionists	Intuitionists argue that judgements about value do not need justifying or proving but that they are self-evident to a person who directs their mind towards moral issues.
Consequentialism or utilitarianism	Consequentialism or utilitarianism bases value decisions on the consequences of an action and not on the actions themselves. So an action that produces the greatest good or 'utility' for the greatest number of people holds the greatest value.
Non-consequentialism or deontological ethics	Non-consequentialism or deontological ethics is concerned with the intentions of managers and not with the consequences of their actions. In other words, an action holds value if it was intended to do good whether or not good actually comes of it in practice. This is assuming that the action comes from a sense of duty to others rather than self-interest. Associated with the writings of Immanuel Kant, deontological (duty-based) theories are about 'doing the right thing' – a slogan we often hear from corporate social responsibility managers in the construction industry (Loosemore and Phua 2011).
Virtue ethics	Virtue ethics emphasises the character of the change agent, rather than their intentions or consequences. It argues that an action holds value if it is one that a virtuous person would do in the same circumstances. The roots of virtue ethics lie in ancient Greek philosophers such as Plato and Aristotle.
Situation ethics	Situation ethics rejects prescriptive rules around value and argues that value judgements should be context-dependent and take account of various constraints, assuming that the underlying intention was to seek the best for the people involved.
Ideology ethics	Ideology ethics argues that ethics is the codification of political ideology and that actions hold no value if they are used by the dominant political elite to control others, particularly if they do not apply this code to themselves.

judgements about a social programme/initiative. Table 1.1 also shows that even a deep understanding of ethics is not likely to provide a definitive judgement of social value, but only serves to provide us with a set of general principles which we can use to do so. Indeed, most philosophers would argue that all ethics can do is eliminate confusion and clarify the issues surrounding a value judgement and that ultimately it is up to each individual to arrive at their own conclusions.

Chapter 3 discusses the complex field of ethics and social value in detail; here we trace how the field of ethics provides many ways to arrive at a determination of social value, and how the specific study of value, as a concept in itself, has evolved in the field of axiology and through the formative work of philosophers such as Nicolai Hartmann – one of the dominant German philosophers of the early twentieth century. This work in complex but in simple terms, axiologists question the simplicity of ethically based notions of value by arguing that moral value (whether something is right or wrong) is only one aspect of goodness and that there are other types of value which should be considered in making any value judgement. These include aesthetic value (capacity to elicit pleasure), and prudential value (capacity to deliver wellbeing and happiness). The field of axiology also asserts that the property of goodness can either be non-natural/intuitive (have properties not discoverable or quantifiable by science or by human sense) or naturalistic (have properties that can be described by physical science). In other words, some aspects of value can be easily measured, but others cannot. We discuss the process and challenges in measuring social value in Chapter 5. However, as Mulgan (2010) pointed out in his formative article on measuring social value, the main obstacle in understanding social value has been the assumption by economists that it is objective, fixed, stable and measurable and responds in predictable ways to interventions. In contrast, social value is a subjective concept where the cause and effects of social interventions are much less clear than those represented by economic theory and more complex to predict because they affect people who attribute different measurements of value depending on their beliefs, ethics, morals, values and priorities. As Mulgan (2010) notes, societies comprise many different and often conflicting systems of social valuation, and that attempts to measure social value have too often ignored this. Similarly, Harlock (2013) argues that attempts to transfer practices from business and economics to social value assessment, in an attempt to give it some scientific credibility, have failed because they assume that social value behaves in the same way as economic value. For example, while many economists would assume that value depreciates over time and apply discounting methods to reflect this, social value often increases over time due to multiplier effects into wider communities and future generations. Similarly, estimations of future value compared to today's value are also complicated by moral arguments about the relative needs of future generations compared to the needs of current generations. Social value also varies across 'time, people, places and situations' (Mulgan 2010: 4) and as Social Value International (2016) states, some but not all of this value is captured in market prices and is often not expressed or measured in the same way as financial value.

This brings us to other aspects of the debates in axiology, which are critical in fully understanding the concept of social value and how it is best articulated and measured. For example, theories of value can also be classified as 'subjectivist' or 'objectivist'. Subjectivist theories argue that judgements of value wholly depend on the subjective judgements of human beings, while objectivists argue that something can have inherent value independently of human judgement. For example, an objectivist would argue an environmental initiative by a construction company which helps to avoid the extinction of a rare species can be legitimately seen as good in itself, regardless of whether people judge it to be good or not. Asking people what they think is good or not, is therefore not always the best way to judge its value. A subjectivist would argue the opposite – that all value is in the eye of the beholder and that any assessment of value should consider all perspectives.

These ideas are related closely to another important distinction in value theory which is the one some people draw between things valued as 'means' (instrumental or extrinsic value) or 'ends' (non-derivative or intrinsic value). The concept of non-derivative or intrinsic value is based on the idea that something can be good for its *own* sake. For example, an area of untouched wildernesses has intrinsic value whether one appreciates it or not. In contrast, something has instrumental/extrinsic value not for its own sake, but because it leads to something which has intrinsic value (it is a means to an end). For example, some may argue that providing an unemployed person with income through employment on a construction project does not have intrinsic value in itself, a point that is debatable as employment has a positive psychological and wellbeing impact in and of itself, but is instead of instrumental/extrinsic value because of the enjoyment derived from the services or commodities that person can now buy. This distinction raises interesting questions for assessing social value because it highlights the potential danger of double-counting if one measured every single instrumental/extrinsic benefit on the path to the ultimate intrinsic benefit of a social intervention. It is for this reason that one of the first steps in undertaking a social impact assessment is to articulate a 'Theory of Change' which describes in one simple diagram the value chain between intervention and impact which diagrammatically links an organisation/project/intervention's inputs, activities, outputs, outcomes and impacts on the stakeholders which are affected. We will discuss how to create a theory of change in Chapter 5.

The complexities of measuring social value become even more apparent when we consider the 'principle of organic unities' which argues that the value of a whole cannot be assumed to be the same as the sum of its parts (Moore 1903). For example, while a painting may hold enormous intrinsic value, the value of the canvas and paint and even time used to make it, may hold very little value. How one considers this in measuring social value has not yet been resolved. The problem with most social impact studies is that they simply add up the combined values for all stakeholders (after accounting for counterfactuals).

Finally, theories of value raise important questions that go beyond simply *whether* something is good or bad to include *how* good or bad it is. In practical terms, this raises questions about how we weight the value judgements of specific elements of extrinsic value produced for different stakeholders affected by a construction project. How does one decide whether one stakeholder's opinion is more important than another's is? There are also probabilities to consider in measuring social value since deterministic estimates of value are unlikely to reflect the stochastic nature of the real world of value (that most estimates of social value have a range of possible outcomes depending on certain events happening). The notion of social value is not static and may also change over time. Indeed, the 'particularist' branch of philosophy argues that because the intrinsic value of something may vary from context to context, a project's value may polarity from good to bad, or *vice versa* (Dancy 2000).

Given the above, we agree with Harlock (2013) who warns that social impact measurement evaluations should never be taken as precise. While many may create the illusion of being scientific from the outside, there are a whole range of opportunities for discretion in the evaluation process which can bias results, from who carries it out, to the selection and identification of indicators, to deciding which stakeholders to consult and involve, to deciding what data is collected and by which methods, and finally to the analysis and presentation of results where there are often strong incentives for organisations to inflate impacts or to be selective in presenting their results. Chapter 5 presents an approach to social impact measurement developed to minimise these risks.

Barriers to delivering social value

There appears to be an unprecedented opportunity for the built environment to leave a more positive legacy in the communities in which it operates. However, recent research reveals a slow uptake of the ideals and practices on social value and a whole array of barriers which currently prevent this from happening (see for example Loosemore and Higgon 2015; Loosemore 2016; Burke and King 2016; Farag, McDermott and Huelin 2016; Petersen and Kadefors 2016; Reid and Loosemore 2017; Denny-Smith and Loosemore 2017). For example, Burke and King (2015) found that while the Social Value Act in the UK had attracted widespread support and interest from both the public and private sector, adoption by local authorities had been inconsistent and up to 75 percent have not included a reference to the Act or social value in their corporate procurement strategies. Furthermore, small- to medium-sized businesses in the UK construction sector, which are often the organisations left to actually deliver social value outcomes on the ground, tend to be unaware of the Act and have little visibility of the wider social value picture they are helping to create.

Farag, McDermott and Huelin (2016) also found that the delivery of social value outcomes in the UK is currently surrounded by confusion in publicly procured construction projects due to a fundamental conflict between public clients' and private contractors' time-horizons, perspectives and motivations in delivering social value. According to Farag, McDermott and Huelin (2016) public clients' perceptions of social value are typically grounded in visions of long-term impact, while contractors have a shorter-term, transactional, project-based perspective [of social value and otherwise] that reflects the fact they have to respond to constantly shifting social priorities on projects as they move from one community to another. It is the temporary, transitionary and constantly shifting context of construction activity that differentiates the notion of social value creation in construction from other contexts; a challenge that has been largely neglected in the wider social value literature. Farag, McDermott and Huelin (2016) also point to a lack of practical guidance on social value delivery processes and confusion among public procurers as to how they reconcile social value objectives with seemingly conflicting procurement rules, laws and regulations designed to ensure probity, value for money and competitive neutrality. They point out that public procurers often get confused by the increasing array of unfamiliar social value outcomes which they are required to deliver, tending to justify their social value decisions retrospectively because of their soft non-quantifiable nature which conflicts with the prevailing procurement culture of quantifying outcomes such as cost and time. Clearly, a focused, pragmatic and legally conversant approach, as detailed in this book, will help more practitioners create effective approaches that provide demonstrable positive impact.

In Sweden, Petersen and Kadefors (2016) report that although some municipalities, and public and private clients, now regularly include employment-creating requirements in their tender requirements when procuring construction services, many aspects relating to social criteria procedures in procurement processes remain unclear. In Australia, Reid and Loosemore (2017) found that social procurement in large contractors is currently compliance-driven, confined to low value and low risk activities, and delivered mainly by existing industry incumbents who do not have the skills, knowledge and ability to collaborate with the third sector to deliver social value. Finally, by studying the implementation of social procurement and the activities of social enterprises across a range of a country's construction industries, Loosemore (2016) highlights a whole array of organisational, procedural, institutional and psychological barriers to the effective integration of third sector organisations into the construction industry. These barriers include:

- Confusion and lack of understanding and guidance about what social value and social procurement means.
- Not having the knowledge and skills and experience of managing and integrating the competing institutional logics of the government, business, third and community sectors in achieving collaborative social outcomes.
- Negative perceptions of third sector organisations (often from bad past experiences) colouring procurement decisions.
- Low competitiveness of third sector organisations due to poor management skills, low reliability, lack of experience in the construction industry, higher price and lower quality compared to normal subcontractors/suppliers.
- Concerns around conflicting objectives and third sector organisations always putting social before commercial objectives.
- Construction firms not taking CSR seriously and competing priorities and imperatives between head office and site.
- Cosy existing supply chain relationships which create strong path-dependencies in procurement decision-making.
- Complex procurement practices and red tape which make it difficult for third sector organisations to tender for construction work.
- The exclusion of third sector organisations from informal networks in getting jobs.
- Resistance and counter-lobbying from industry incumbents which are likely to lose work to new third sector entrants.

- The highly regulated nature of construction around issues like safety and the need for certifications which many third sector organisations do not have.
- A lack of evidence that third sector organisations deliver higher social value than existing suppliers.
- A lack of imagination of how social procurement and third sector organisations can add value in bids by major contractors.
- Third sector organisations not having the resources and skills to report and communicate the social value they create.

Whilst it is discouraging to know that these formidable barriers exist in the built environment, it is important to note that these are not limited to construction; similar barriers have been identified in other sectors (Kernot and McNeill 2011).

Summary

The aim of this chapter was to introduce the concept of social value within the context of the built environment, showing that the introduction of this term into the sector's lexicon reflects wider and fundamental changes in public governance, welfare provision and societal expectations which in turn are altering the relationship between the sector and the community in which it operates. While many leading firms in the industry have a long and proud tradition of contributing to the communities in which they build, the vast majority of professionals who work in the sector have little understanding of what the term social value means, why it has emerged, how to create it and how it will affect future roles, practices and relationships within the industry.

Key points

- Social value is simply the 'social impact' any construction organisation, project or programme makes to the lives of internal and external stakeholders affected by its activities, including those working in the industry and in the communities in which it operates.
- Interest in social value is being driven by wider trends in CSR, new public governance and social procurement.
- Social value can be created at all stages of a construction project from urban planning through design, tendering, construction and operations and facilities management.
- There is no one best way to create social value. This depends on the maturity, experience and resource constraints of a business and a strategic approach needs to be taken driven bottom-up by the needs of the community.
- No one organisation alone can create social value. Collaboration is key – working with different organisations from across the community, public, private, third and not-for-profit sectors.
- The potential commercial benefits of showing leadership in creating social value are enormous.

References

ABS (2002) *1301.0 – Year Book Australia, 2002*. Australian Bureau of Statistics: Canberra, Australia.

ABS (2017) *6202.0 – Labour Force, Australia, May 2017*. Australian Bureau of Statistics: Canberra, Australia.

Anderson, E. (1993) *Value in Ethics and Economics*. Harvard University Press: Cambridge, MA.

Anderson, J., Ruggeri, K., Steemers, K., and Huppert, F. (2016) Lively social space, well-being activity, and urban design: findings from a low-cost community-led public space intervention. Environment and Behaviour, 49: 685–716.

Aubusson, K. (2016) Locals chain themselves to trees that face the chop for Sydney light rail in Randwick. *Sydney Morning Herald* [online]. Available at: www.smh.com.au/nsw/locals-chain-themselves-to-trees-that-face-the-chop-for-sydney-light-rail-in-randwick-20160107-gm1ie8.html [accessed 17th June, 2016].

Australian Council of Superannuation Investors (2017) *Corporate Sustainability Reporting*. Australian Council of Superannuation Investors: Sydney, Australia.

Awale, R. and Rowlinson, S. (2014) A conceptual framework for achieving firm competitiveness in construction: A 'creating shared value' (CSV) concept. In: Raiden, A. and Aboagye-Nimo, E. (eds), *Proceedings 30th Annual ARCOM Conference, 1–3 September 2014*, Association of Researchers in Construction Management: Portsmouth, UK, pp. 1285–1294.

Barnes, J. (1995) *'Life and Work'. The Cambridge Companion to Aristotle*. Cambridge University Press: Cambridge.

Barraket, J., Keast, R., and Furneaux, C. (2016) *Social Procurement and New Public Governance*. Routledge: London.

Barraket, J. and Loosemore, M. (2018) Co-creating social value through cross sector collaboration between social enterprises and the construction industry. *Construction Management and Economics*, 36(7): 394–408.

Barraket, J. and Weissman, J. (2009) *Social Procurement and its Implications for Social Enterprise: A Literature Review*. The Australian Centre for Philanthropy and Nonprofit Studies, Queensland University of Technology: Brisbane, Australia.

Bartlett, B., Dalgleish, J., Mao, S., and Roberts, E. (2012) *Reconnecting Disaffected Youth Through Successful Transition to Work*. Griffiths University: Queensland, Australia.

Barton, H., Thompson, S., and Grant, M. (2015) Urban Inequities, Population Health and Spatial Planning. *The Routledge Handbook of Planning for Health and Well-being: Shaping a Sustainable and Healthy Future*. Routledge: London.

Benington, J. and Moore, M.H. (2011) Public value in complex and changing times. In: Benington, J. and Moore, M.H. (eds), *Public Value: Theory and Practice*, Palgrave: London, pp. 1–20.

Beschorner, T. (2013) Creating shared value: the one-trick pony approach – a comment on Michael Porter and Mark Kramer. Business Ethics Journal Review, 17(1): 106–112.

Blin, S. and Eldred, A. (2016) Apprenticeships, Crossrail Learning Legacy: Case Study. Crossrail Lerning Legacy. Available at: http://learninglegacy.crossrail.co.uk/documents/apprenticeships/ [accessed 14th September, 2017].

Bonwick, M. and Daniels, M. (2014) *Corporate Social Procurement in Australia: Business Creating Social Value*. Social Traders: Melbourne, Australia.

Brouwers J., Prins, E., and Salverda M. (2010) *Social Return on Investment: A Practical Guide for the Development Cooperation Sector. Context, International Cooperation.* Context, International Cooperation: Utrecht, The Netherlands.

Brown, M. (2005) *Corporate Integrity: Rethinking Organizational Ethics and Leadership.* Cambridge University Press: Cambridge.

Burke, C. and King, A. (2015) *Generating Social Value through Public Sector Construction Procurement: A Study of Local Authorities and SMEs.* School of Architecture, Design and the Built Environment, Nottingham Trent University: Nottingham, UK.

Burkett, I. (2010). *Social Procurement in Australia (Including A Compendium of Case Studies).* Foresters Community Finance, The Centre for Social Impact: University of NSW: Sydney, Australia.

Cabinet Office (2012) *A Guide to Social Return on Investment.* Cabinet Office: London.

Cabinet Office (2015) *Social Value Act Review – Report February 2015.* Cabinet Office: London.

CICA (2015) *2015 Construction Industry Outlook.* Career Industry Council of Australia: Sydney, Australia.

City of Toronto (2014) *Social Procurement Framework.* Toronto, Canada.

Clement, L., Damien, P., Shabnam, A., Martinez, E., and Thaler, P. (2010) *A Quantitative Analysis of Value Drivers Associated with Sustainable Procurement Practices.* PwC and EcoVadis in collaboration with the INSEAD Social Innovation Centre.

Commission for Architecture & the Built Environment (CABE) (2002) *The Value of Good Design: How Buildings and Spaces Create Economic and Social Value.* Commission for Architecture and the Built Environment: London.

Commonwealth of Australia (2015a) *2015 Intergenerational Report Australia.* Commonwealth of Australia: Canberra, Australia.

Commonwealth of Australia (2015b) *Commonwealth Indigenous Procurement Policy, July 2015.* Commonwealth of Australia: Canberra, Australia.

Commonwealth of Australia (2017) *Closing the Gap Prime Minister's Report 2017.* Australian Government, Department of Prime Minister and Cabinet, Commonwealth of Australia: Canberra, Australia.

Cook, M. and Monk, G. (2012) *The Social Value Guide: Implementing the Public Services (Social Value) Act.* Social Enterprise UK: London.

Corburn, J. (2015) Urban Inequities, Population Health and Spatial Planning. In: Barton, H., Thompson, S., Burgess, S., and Grant, M. (eds) *The Routledge Handbook of Planning for Health and Well-being: Shaping a Sustainable and Healthy Future.* Routledge: London, pp. 37–47.

Crane, A., Palazzo, G., Spence, L.J., and Matten, D. (2014) Contesting the value of 'creating shared value'. *California Management Review,* 56(2): 130–154.

Croydon Council (2012) *Inspiring and Creating Social Value in Croydon: A Social Value Toolkit for Commissioners.* Croydon Council: London.

CSI (2014) *The Compass: Your Guide to Social Impact Measurement.* Centre for Social Impact: Sydney, Australia.

Dancy, J. (2000) The Particularist's Progress, In: Hooker, B. and Little, M. (eds), Moral Particularism. Clarendon Press: Oxford, pp. 130–156.

Denny-Smith, G. and Loosemore, M. (2017) Integrating Indigenous enterprises into the Australian construction industry. *Engineering, Construction and Architectural Management,* 24(5): 788–808.

Doherty, B., Haugh, H., and Lyon, F. (2014) Social Enterprises as hybrid organizations: a review and research agenda. *International Journal of Management Reviews,* 16(4): 417–436.

EC (2010) *Buying Social: A Guide To Taking Account of Social Consideration in Public Procurement*. European Commission, Publications Office of the European Union: Luxembourg.

Emerson, J. (2000) *The Nature of Returns: A Social Capital Markets Inquiry into Elements of Investment and the Blended Value Proposition*. Social Enterprise Series No. 17, Harvard Business School: Boston, MA.

Ernst & Young (2012) *Managing Bribery and Corruption Risks in the Construction and Infrastructure Industry*. Ernst & Young: London.

Esteves, A. M. and Barclay, M. (2011) Enhancing the benefits of local content: integrating social and economic impact assessment into procurement strategies. *Impact Assessment and Project Appraisal*, 29(3): 205–215.

Farag, F., McDermott, P. and Huelin, C. (2016) The Development of an Activity Zone Conceptual Framework to Improve Social Value Implementation in Construction Projects Using Human Activity Systems. School of the Built Environment, The University of Salford.

Farmer, B. (2016) *The Farmer Review of the UK Construction Labour Model*. Construction Leadership Council: London.

Fleming, R.L. (2007) *The Art of Placemaking: Interpreting Community Through Public Art and Urban Design*. Merrell Publishers: New York.

Furneaux, C. and Barraket, J. (2014) Purchasing social good(s): a definition and typology of social procurement. *Public Money & Management*, 34(4): 265–272.

G8 (2014) *Measuring Impact: Guidelines for Good Impact Practice*. Impact Measurement Working Group of the Social Impact Investment Taskforce (established by the G8): London.

GCP (2013) *Global Construction 2025*. Oxford Economics, Global Construction Perspectives Ltd: London.

GECES (2014) *Proposed Approaches to Social Impact Measurement in European Commission Legislation and in Practice Relation to: EuSEFs and the EaSi*. GECES Sub Group on Impact Measurement: Paris.

Griffith, A., Knight, A., and King, A. (2003) *Best Practice Tendering for Design and Build Projects*. Thomas Telford: London.

Grisolia, F. and Ferragina, E. (2015) Social innovation on the rise: yet another buzzword in a time of austerity? *Salute e società*, 1: 169–179.

Grob, S.M. and Benn, S.H. (2014) Conceptualising the adoption of sustainable procurement: an institutional theory perspective. *Australasian Journal of Environmental Management*, 21(1): 11–21.

HACT (2016) *Social Value and Procurement: A Toolkit for Housing Providers and Contractors*. Housing Associations Charitable Trust: London.

Harlock, J. (2013) Impact measurement practice in the UK third sector: a review of emerging evidence [Working paper: full academic study]. Third Sector Research Centre, Working Paper 106: Birmingham, UK. Available at: www.construction-manager.co.uk/management/opening-our-eyes-supply-chain-ethics/ [accessed 4th February, 2015].

Horrigan, D. (2011) *Strategic Serendipity: The Art of Being in the Right Place at the Right Time ... with the Right People*. Australia Business Foundation: Sydney, Australia.

ICAEW (2015) *Quantifying Natural and Social Capital: Guidelines on Valuing the Invaluable*. Institute of Chartered Accountants in England and Wales: London.

International Organisation for Standardisation (ISO) (2017) *ISO 20400 Sustainable Procurement – Guidance*. International Organization for Standardisation. Geneva, Switzerland.

International Organisation for Standardisation (ISO) (2010) *ISO 26000 Social Responsibility*. International Organisation for Standardisation: Genève Switzerland.

Indigenous Procurement Policy (2015) *Indigenous Procurement Policy*. Australian Government: Canberra, Australia.

Jenkins, S.P. (2015) *The Income Distribution in the UK: A Picture of Advantage and Disadvantage*. Institute of Social and Economic Research: London.

Johnston, C. (1992) *What is Social Value?* Australian Government Publishing Service: Canberra, Australia.

Jones, T.M. (1995) Instrumental stakeholder theory: a synthesis of ethics and economics. *The Academy of Management Review*, 20(2): 404–437.

Kanter, R.M. (1999) From spare change to real change: the social sector as beta site for business innovation. *Harvard Business Review*, 77(3) (May/June): 122–128.

Kelly, J., Male, S., and Graham, D. (2007) *Value Management of Construction Projects*. Blackwell: Oxford.

Kernot, C. and McNeill, J. (2011) *Australian Stories of Social Enterprise*. University of New South Wales: Sydney, Australia.

Kritkausky, R. and Schmidt, C. (2011) *Handbook for Implementers of ISO 26000, Global Guidance Standard on Social Responsibility*. London: ECOLOGIA.

LePage, D. (2014) *Exploring Social Procurement*. Accelerating Social Impact CCC Ltd: Vancouver, BC, Canada.

Loosemore, M. (2016) Social Procurement in UK Construction Projects. *International Journal of Project Management*, 34(1): 133–144.

Loosemore, M., Raftery J., Reilly, C., and Higgon D. (2005) *Risk Management in Projects*, 2nd edn. Spon Press: London.

Loosemore, M. and Higgon, D. (2015) *Social Enterprise in the Construction Industry: Building Better Communities*. Routledge: London.

Loosemore, M. and Lim, B. (2017) Linking corporate social responsibility and organisational performance in the construction industry. *Construction Management and Economics*, 35(3): 95–105.

Loosemore, M. and Phua, F. (2011) *Corporate Social Responsibility in the Construction Industry: Doing the Right Thing?* Routledge: London.

Macmillan, R. (2013) Decoupling the state and the third sector? The 'Big Society' as a spontaneous order. Third Sector Research Centre, Working Paper, 101. University of Birmingham: Birmingham, UK.

Maier, F. and Schober, C.S., Simsa, R., and Millner, R. (2015) SROI as a method for evaluation research: understanding merits and limitations. *Voluntas*, 26(5): 1805–1830.

Mandell, M., Keast, R., and Chamberlain, D. (2016) Collaborative networks and the need for a new management language. *Public Management Review*, 19(3): 326–341.

Masom, D., Martyres, R., and O'Brien, R., supported by Markey, K. (2013) *The Landmark Difference*. Social Enterprise UK: London.

Master Builders Association (2017) *Towards 2020 Policy for Australian Apprenticeship Reforms*. 2025 Master Builders Association: Melbourne, Australia.

McCrudden, C. (2004) Using public procurement to achieve social outcomes. *Natural Resources Forum*, 28(4): 257–267.

Mitchell, H. (2012) Fostering creative communities key to escaping limiting island mindset. The Sydney Morning Herald, 10 November, Weekend Business, 8–9.

Moore, G.E. (1903) *Principia Ethica*. Cambridge University Press: Cambridge.

Moore, M.H. (1995) *Creating Public Value*. Harvard University Press: Cambridge, MA.

Mulgan, G. (2010) Measuring social value. Stanford social innovation review. Summer 2010. Available at: https://ssir.org/articles/entry/measuring_social_value [accessed 20th June, 2016].

NCVO (2013) *Inspiring Impact: The Code of Good Impact Practice*. National Council for Voluntary Organisations, London: London.

Newman, C. and Burkett, I. (2012) *Social Procurement in NSW – A Guide to Achieving Social Value through Public Procurement*. Social Procurement Action Group: Sydney, Australia.

Nicholls, J., Sacks, J., and Walsham, M. (2005) *A Guide to Procuring for Social Enterprises: More For Your Money*. Social Enterprise Coalition: London.

NPC (2014) *Building Your Measurement Framework: NPC's Four Pillar Approach*. New Philanthropy Capital: London.

NSW (2015) *Principles for Social Impact Investment Proposals to the NSW Government, Office of Social Impact Investment*. NSW Government: Canberra, Australia.

OECD (2017) *Understanding the Socio-Economic Divide in Europe Background Report 26 January 2017*. OECD: Paris.

Perry, R.B. (1967) *General Theory of Value: Its Meaning and Basic Principles Construed in Terms of Interest*. Harvard University Press: Cambridge.

Petersen, D. and Kadefors, A. (2016) *Social Procurement and Employment Requirements in Construction*. KTH Royal Institute of Technology: Stockholm.

Porter, M.E. and Kramer, M.R. (2006) Strategy and society: the link between competitive advantage and corporate social responsibility. *Harvard Business Review*, December, 2006: 78–92.

Porter, M.E. and Kramer, M.R. (2011) Creating shared value. *Harvard Business Review*, 89 (January/February 2011): 62–77.

Preferential Procurement Policy Framework Act (2000) *Preferential Procurement Policy Framework Act 5, Number 5*, Government of South Africa: Johannesburg, South Africa.

Preferential Procurement Regulations Act (2017) Government of South Africa: Johannesburg, South Africa.

Procurement Strategy for Aboriginal Business (PSAB) (2016) *Procurement Strategy for Aboriginal Business (PSAB), Aboriginal Affairs and Northern Development Canada (AANDC)*. Government of Canada: Quebec, Canada.

Public Services (Social Value) Act (2012) UK Cabinet Office: London.

Reid, S. and Loosemore, M. (2018) The social procurement practices of tier-one construction contractors in Australia. Construction Management and Economics, In Press.

RICS (2011) *Written Evidence Submitted by RICS (LOCO 105)*. Localism – Communities and Local Government Committee, UK Parliament: UK. Available at: https://publications.parliament.uk/pa/cm201012/cmselect/cmcomloc/547/547dvw19.htm [accessed 1st April, 2016].

Schneewind, J.B, Baron, M., Kagan, S., and Wood, A.W. (2002) *Groundwork for the Metaphysics of Morals, Immanuel Kant*, A.W. Wood (Ed. and Trans.). Yale University Press: New Haven, CT.

Social Enterprise UK (2012) *The Social Value Guide: Implementing the Public Services (Social Value) Act*. Social Enterprise UK: London.

Social Value International (2016) *Understanding and Use of the Principles of Social Value*. Social Value International: London.

Supply Chain Sustainability School (2018) *Modern Slavery in Australian Property, Construction and Infrastructure Supply Chains: Understanding Risks and Relevance*. Supply Chain Sustainability School: Sydney, Australia.

The Public Law 95–507 Act (1978) Public Law 95–507 Act – OCT. 24 (1978) United States of America Government: Washington, DC.

Transparency International (2011) *2011 Bribe Payers Index Report.* Transparency International: Berlin.

UK Cabinet Office (2012) *A Guide to Social Return on Investment.* UK Cabinet Office, The Office of the Third Sector: London.

Vogelius, P. and Storgaard, K. (2016) Cooperation in Construction: The Role of Values. In: Chan, P.W. and Neilson, C.J. (eds), *Proceedings 32nd Annual ARCOM Conference*, 5–7 September 2016, Association of Researchers in Construction Management: Manchester, UK, pp. 135–144.

Watts, G., Dainty, A., and Fernie, S. (2015) *Making Sense of CSR in Construction: Do Contractor and Client Perceptions Align?* School of Civil and Building Engineering Loughborough University: Loughborough, UK.

Watts, P. and Holme, L. (2003) *Corporate Social Responsibility.* World Business Council for Sustainable Development: Geneva.

Whelan, J. (2012) *Big Society and Australia.* Centre for Policy Development, May 2012: Melbourne, Australia.

Wiggins, D. (1987) *Needs, Values, Truth.* Blackwell: New York.

William, T. (2009) *The Search for Leadership: An Organisational Perspective.* Triarchy Press: Bridport, UK.

Williams, R. and Hayes, J. (2013) Literature Review: Seminal Papers on 'Shared Value'. Economic and Private Sector Professional Evidence and Applied Knowledge Services, available at http://partnerplatform.org/?5mar03m5 [accessed 17th March, 2017].

WMI (2010) *The Future of Global Construction 2014. Market Intelligence Report.* World Market Intelligence: London.

Yourtown (2017) *'I've got a place' Indigenous Participation Strategies – Practice Insights with Applications for the Construction Industry.* Yourtown: Brisbane, Australia.

2 Social value in the built environment

The legal framework in the UK

Beulah Allaway and Martin Brown,
Anthony Collins Solicitors LLP

This chapter discusses the legal framework in the UK surrounding social value in contracts linked to the built environment, particularly referring to contracts for traditional construction works and infrastructure projects. In such contracts, there are significant opportunities to secure social value, particularly within infrastructure projects.

Traditionally, the construction sector has been at the forefront of achieving social value through its contracts, driven by both public sector employers understanding that they can maximise community benefit through how they procure, in addition to contractors themselves understanding the commercial benefits of training, employment and environmental schemes. The introduction of a legal framework has given an opportunity to galvanise this sector experience to achieve even greater social value through public works contracts. This is because the legal framework places duties on public sector employers to consider social value.

The legal framework for social value creation

The legal framework surrounding social value relates to public sector employers in England and Wales, who must balance a duty to consider social value under the Public Services (Social Value) Act 2012 ('the Social Value Act') with strict compliance of tendering rules set out in the Public Contracts Regulations 2015 ('the Public Contracts Regulations').

Given the nature of those affected by this legal framework, whenever reference in this chapter is made to an 'employer' under a contract, this should be read as an employer that is a 'contracting authority' subject to the Public Contracts Regulations and/or the Social Value Act, and to contracts with a value above the relevant tendering threshold in the Public Contracts Regulations that requires procurement in accordance with one of its prescribed procedures. Whilst private sector employers (or public sector ones where contracts are below such thresholds) are not bound by the same legal obligations, the principles described in this chapter can still be applied proportionately to effectively deliver social value.

Social value is an exciting area of public procurement law, which can deliver increased efficiencies, better value for money and social benefits in an economic climate that continues to be challenging. With increased budgetary pressure, public sector employers may be tempted to focus more on price and less on value for money. However, getting the most out of public expenditure is even more important when finances are tight and economic growth is restrained.

In the context of Brexit, the UK's departure from the European Union ('the EU'), there may be a temptation to disengage with the public procurement rules and social value. However, the Social Value Act is deliberately non-prescriptive: its goal is to give public sector employers the opportunity to engage with social value, not to strong-arm them into doing so. It is also domestic law and so, in principle, will not be affected by Brexit. Furthermore, the European Union (Withdrawal) Bill 2017–2019[1] will transpose the entirety of EU law into UK law on the day the UK's exit from the EU takes place. This means that, in the short term at least, the Public Contracts Regulations originating in EU law, will also survive Brexit. It is impossible to foresee the ultimate impact of Brexit for public procurement law, but post-EU referendum government publications discussed below, demonstrate a continued commitment to social value albeit that the government's terminology seems to be changing.

Despite the opportunities, and a legal framework and government policy that supports them, challenges continue to result in employers failing to maximise social value when tendering. These challenges are interrelated and primarily are:

- the tension between the need to prescribe what is being procured, and the desire to free-up contractors to innovate;
- a lack of understanding of what social value is and underappreciating what can be achieved;
- inconsistent approaches between procurement and contract management; and
- procuring social value which is not linked to the subject matter of the contract.

This chapter addresses these challenges by exploring where the greatest opportunity for social value through public procurement appears to be at present; setting out how public policy supports such conclusions; explaining the opportunities and restrictions presented by the legal framework; and providing a practical guide for employers on how to consider and address social value at every stage of procurement in order to ensure that opportunities identified are maximised.

Understanding social value – opportunities in the built environment

A lack of understanding of what social value is and underappreciating what can be achieved can inhibit its impact, as it contributes to a view that social value is akin to 'added value'. Social value should not be limited to added value that can be bolted onto the back of a contract. So much more can be achieved from social value than this: Social value can and should be about rethinking a contract, and how it is procured, to maximise social benefits.

Within the construction sector there are significant opportunities for employers to secure social value. Infrastructure projects are a good example, each representing a hub of such opportunity. This is because most infrastructure projects are not contracts being re-let, but are entirely new contracting opportunities. This means that they typically involve a new workforce, rather than a workforce transferred under the Transfer of Undertakings (Protection of Employment) Regulations 2006 ('TUPE'), so there is huge scope to require the provision of training and introduce new entrants to the job market. In addition, they are, by their nature, often seen as a public sector social and economic investment, looking for innovation to achieve return on the high-value investment, rather than allowing how things have previously been done to constrain the new contracting arrangements.

As a consequence of these factors, there is much scope for social value to be creatively considered at the conception stage of an infrastructure contract as a key objective to its delivery. In effect, such projects provide a blank slate for considering social as well as economic benefits as central contracting objectives.

With Brexit approaching and the likelihood of substantial changes to the freedom of movement, the UK's ongoing reliance on foreign labour in the construction sector is increasingly challenging in a time of critical skills shortages in the domestic labour market. One key area in which social value can be achieved in infrastructure is recruitment of, and training for, new entrants. The length of infrastructure projects (often taking months and sometimes years), and the construction 'hub' created, make them particularly well suited to training schemes, such as apprenticeships and graduate schemes. Infrastructure maintenance works, due to their long-term contracts, are also well suited to securing the same, albeit that the implications of TUPE will more often than not have an impact on the workforce.

Recent government publications support an agenda for maximising social value opportunities through public contracts, particularly public infrastructure projects. For example, social, economic and environmental factors are a strong theme in the Cabinet Office's October 2016 'Procurement Policy Note – Procuring for Growth Balanced Scorecard',[2] which discusses social benefits in a range of contexts (supply chain accessibility for SMEs, employment of disadvantaged groups and training opportunities, for example).

This policy note makes it mandatory for projects valued over £10 million to use the Balanced Scorecard tool to balance tender scoring between straightforward matters such as cost and more complex issues (such as social value and wider socio-economic considerations). The use of the Balanced Scorecard approach can aid employers to secure social value through tenders in a way that tackles 'head on' the tension in social value between the need to prescribe what is being procured and the need to facilitate innovation.

In addition, the 'Building our Industrial Strategy' Green Paper published in January 2017,[3] and the 'Industrial Strategy' White Paper[4] that followed in November 2017, focus on social value and relevant issues, albeit without many specific references to the phrase 'social value'. The Industrial Strategy is ambitious and suggests that the government's focus shall be on a major upgrade to the UK's infrastructure. Where borne through, this major upgrade, taken together with the government's focus on partnering with private business to achieve its infrastructure goals, represents one of the largest opportunities ahead to secure social value through public works contracts.

The Local Government Association's 2018 edition of 'National Construction Category Strategy'[5] discusses social value in the context of construction. It highlights a focus on employment and skills and aims to achieve increased social value through standardised selection questionnaires (aided by the recently updated selection questionnaire that must be used when tendering public works contracts, PAS 91). It also advocates advertising local supply chain opportunities and making fair payment to subcontractors. This strategy is more direct than central government publications in explicitly naming social value as an objective and setting out some of the practical things that can be done to achieve it.

Most recently, in April 2018, a further Procurement Policy Note 'Supply Chain Visibility' was published focusing on increasing access to SMEs through the public procurement supply chain.[6] This requires all Central Government Departments, their Executive Agencies and Non Departmental Public Bodies in England, from 1 May 2018, to report on how much of their subcontracting spend goes directly to SMEs. In addition these contracting authorities must include terms in contracts valued above £5 million requiring main contractors subject to them to themselves advertise sub-contract opportunities worth over £25,000 on Contracts Finder. Collectively, these post-EU referendum policy publications demonstrate that social value will continue to be a domestic requirement, regardless of our position in or out of the EU.

Understanding social value – the legal framework

Domestic legislation

Understanding the legal framework governing social value is critical to overcoming the challenges employers face when seeking to procure it. This is because it is only by understanding the parameters of what the law allows that opportunities can be truly harnessed.

The duty on employers established by the Social Value Act relates to service contracts above the relevant monetary thresholds in the Public Contracts Regulations. It also applies to contracts with a works/supplies element that is so incidental that the contract would ordinarily be considered a 'services contract' under the Public Contracts Regulations, as well as to framework agreements for the same. Though there are many construction contracts to which the Social Value Act does not apply, the Government is clear that social value should be considered in public procurement of works. It also makes commercial sense for public sector employers to do so as incorporating social value facilitates achieving value for money.

The Social Value Act provides a real impetus by imposing an active duty on relevant employers to consider the economic, environmental and social benefits that can be achieved through commissioning. It does so by requiring consideration by public sector 'authorities' of:

- how what is proposed to be procured might improve the economic, social and environmental well-being of the relevant area; and
- how, in conducting the process of procurement, it might act with a view to securing that improvement.[7]

Public sector authorities are also required to consider whether consultation on social value matters is needed.[8]

In the context of the Social Value Act, authorities subject to this duty are those authorities that are subject to the requirements of the Public Contracts Regulations. The duty applies to the pre-procurement stage,[9] including the need to consider the same when planning to call-off under framework agreements that were first established after the Social Value Act came into force. However, the duty can be overlooked where such consideration would be impractical in a genuinely urgent situation.[10]

The Social Value Act encourages good practice. It does not penalise poor practice. However, the lack of any real penalty for failure to comply with the duty should not limit its impact because applying the duty as designed should mean that authorities find consideration of social value pre-procurement is not an onerous and meaningless tick-box exercise, but is instead core to their procurement processes and policy. Authorities must give genuine consideration to social value and make an active decision about what they procure and how they procure it *before* they procure it. This duty should 'trickle down' to the tender stage itself, for example by ensuring that authorities ask bidders questions that are relevant, pertinent and of genuine policy importance. Where this is not the case, authorities' decisions in relation to social value requirements may be challengeable, for example by unsuccessful bidders (subject to time limits on any challenge – time limits on judicial review, for example, are generally three months from the date the grounds for the review arose, though this is reduced to a time limit of 30 days where the challenge relates to a decision governed by the Public Contracts Regulations).

The Social Value Act forms part of a legislative framework governing how procurements should be run. Other key domestic legislation that is closely related to social value includes the Modern Slavery Act 2015 and Local Government Act 1999. The Modern Slavery Act imposes a requirement for commercial organisations with a turnover of £36 million or more to prepare a slavery and human trafficking statement in each financial year.[11] This statement must include evidence of actions taken to ensure its supply chain is free from slavery, which most often results in a commitment to ensure it is a contract term at all stages of the supply chain that contractors do not commit slavery or human trafficking. The Local Government Act 1999 imposes an obligation on English local authorities to secure best value through continuous improvement in the way its functions are exercised[12] and statutory guidance on the same makes it clear that social value is intrinsically linked with achieving best value.[13] In addition, the Equality Act 2010 prohibits discrimination in relation to a number of key protected characteristics. At section 149 it also imposes the public sector equality duty on public authorities, requiring public authorities to have due regard to the need to eliminate discrimination, advance equality of opportunity and foster good relations.[14]

The legal framework described above applies to England and, mostly, to Wales and Northern Ireland as well, as is explained below.

The Public Contracts Regulations implement Directive 2014/24/EU ('the Consolidated Directive') to apply in Wales and Northern Ireland as they do in England. The Best Value obligations of the Local Government Act 1999, discussed above (requiring public sector authorities to secure best value through continuous improvement) only apply to Wales and England. The Equality Act 2010 contributes to the legal agenda for Wales, Scotland and Northern Ireland, as well as England.

In Wales, the Social Value Act applies to Welsh bodies that are not solely or mainly under the jurisdiction of the Welsh Assembly Government but does not apply to those authorities whose functions are wholly or mainly Welsh devolved functions.[15]

In 2014 the Welsh Government published a guide titled 'Community Benefits Delivering Maximum Value for the Welsh Pound'.[16] Community benefits closely align with social value principles as they aim to address and balance economic, social and environmental issues and impacts. A key approach by the Welsh Government is achieving sustainable development, and achieving community benefits contributes to this goal.[17]

The Social Value Act does not apply in Northern Ireland. This does not mean that employers in Northern Ireland cannot seek to achieve community benefit in their procurement of services (or goods and/or works). In fact, many do already.

Procurement regime and reform in Scotland

Scotland is its own legal jurisdiction and has its own domestic law. This includes the Public Contracts (Scotland) Regulations 2015, which implements the Consolidated Directive in the way that the Public Contracts Regulations do in England, Wales and Northern Ireland.

In addition, the Procurement Reform (Scotland) Act 2014 places general and specific additional duties on employers, including:

- A general duty on regulated authorities to consider the sustainability of their procurement processes before carrying out regulated procurement.[18]
- A specific duty on those authorities anticipating 'significant procurement expenditure' in any coming financial year to draw up a procurement strategy that includes a statement of that authority's policy on the use of community benefit requirements in its contracts.[19]

- A specific duty that such authorities will have to consider whether to include such community benefit requirements as part of the procurement for any contract of a value equal to or greater than £4,000,000.[20]

The Procurement Reform (Scotland) Act goes further than any other piece of legislation adopted in the United Kingdom to date in casting social value at the heart of public procurement. The Act reflects the Scottish culture of embedding community benefits through public procurement.

EU legislation

European Commission guidance and strategy highlight the importance the Commission places on public spending achieving social, economic or environmental goals. Whilst these show a clear desire at Commission level to deliver social value through public procurement, it is important to understand the legislative sources that enable employers to do so in a way which is free from the fear of challenge.

The EU legislative framework leaves us in no doubt that there are instances in which it is valid for a 'contracting authority' to consider social and environmental characteristics when deciding whether to award a contract. This is made clear by Directive 2014/24/EU ('the Consolidated Directive'), which allows contracting authorities to set down 'special conditions relating to the performance of a contract' which 'may include economic, innovation-related, environmental, social or employment-related considerations'.[21]

The Consolidated Directive encourages contracting authorities to engage in social value in the following ways:

- It demonstrates an overarching commitment to social value in EU procurement.[22]
- It aims to foster the involvement of Small and Medium Enterprises ('SMEs') in public procurement, for example by encouraging the division of large contracts into lots.[23]
- Award criteria are intended to assess the value of the tender from the point of view of the contracting authority, emphasising that it is for the contracting authority to choose what it identifies as being of value. Those award criteria can explicitly include social factors.[24]
- Contracting authorities may not award a contract to the otherwise 'best' tender where that tender does not comply with certain social and environmental laws (e.g. International Labour Organization conventions that the UK signs up to).[25]
- It codifies the Dutch Coffee Case[26] that prohibits contracting authorities from requiring bidders to hold a particular label, but permitting them to define the technical specifications for the contract that they require. This allows contracting authorities to specify that certain standards (eg. environmental management systems that are ISO 14001 accredited) are minimum contracting requirements without falling foul of anti-competition practices that could affect trade between member states.[27]

- Social and environmental characteristics can be considered by contracting authorities when determining which tender submitted is the most economically advantageous ('MEAT').[28]
- The contractor's corporate, social and / or environmental policy is not linked with the subject matter of the contract and, as such, cannot be a tendering requirement. Whilst this may at first appear to go against the objectives of the legislation to increase social value in public contracts, in fact the Consolidated Directive here follows existing case law by clarifying that an organisation's corporate policy relates to everything that organisation does; not solely to the subject matter of the contract being tendered and, therefore, is not a legitimate tendering requirement.[29]

The Public Contracts Regulations incorporate the Articles of the Consolidated Directive into domestic law in England, Wales and Northern Ireland. For the purpose of these Regulations 'contracting authorities' means the State, regional or local authorities and bodies governed by public law.

Duties under the Public Contracts Regulations are broader than under the Social Value Act, affecting all public sector services, supplies and works contracts over specified financial thresholds. Under these Regulations, social value can be a legitimate procurement requirement if it:

- is used to assess MEAT in achieving value for money;
- is linked to the subject matter of the contract;[30]
- does not confer an unrestricted freedom of choice on the contracting authority;
- complies with EU Treaty obligations of transparency, equality and proportionately, including in particular that it is not directly or indirectly discriminatory;
- is compatible, generally, with EU law;
- can be compared and/or assessed objectively; and
- is properly advertised in the contract notice (OJEU) and / or in the contract documents.

A failure to understand what these parameters mean in practice, twinned with a fear of procurement challenge, can lead many contracting authorities to assume (incorrectly) that the risk of including social value factors in tender evaluation outweighs the benefits of the social value being procured. Set out below is a brief summary of some of the case law in this area, highlighting a number of instances in which social value requirements have been challengeable. This chapter then sets out a practical guide to applying the principles set by legislation and case law in procurement practice so that contracting authorities can procure social value with confidence.

EU case law

The Consolidated Directive was introduced with the aim of consolidating a great deal of preceding case law from the European Court of Justice ('the ECJ').[31] There have similarly been some significant pieces of case law in the domestic courts.

These cases can sometimes be difficult to reconcile, but collectively they have established important principles, both forming the basis of what is now set out in the Consolidated Directive and the Public Contracts Regulations, and helping us to flesh out and properly understand those principles. It is worth noting that it is not possible in this chapter to explore in detail the extensive case law in this area, nevertheless some of the most impactful principles established in case law are as follows:

- it is possible to include social and environmental requirements in public procurement provided the procurement documents and procurement process do not disadvantage non-local bidders, for example by requiring them to have local labour market knowledge, or a local base, or use local materials (see *Laboratori Bruneau*);[32]
- environmental and (by implication) social requirements that address the purchaser's policy objective are permissible as they do not need to provide an economic benefit to the purchaser (see *Finnish Buses*);[33]
- requirements placed on contractors must be measurable and verifiable (see *EVN*);[34] and
- contractors can be required to use environmentally friendly materials or fair trade products, but can't be asked to provide or carry specific labels or brands (see the *Dutch Coffee Case*),[35] especially when these go beyond what is needed for the specification.

A common thread of such case law is the demonstration that the EU legislative framework embraces social value to the extent that the employer would like to include social value requirements in their procurement. The areas where the law becomes more prescriptive relate to how social value obligations are incorporated in the tender process. For example, social value cannot be sought in a way that disadvantages non-local bidders; social value obligations must be capable of measurement and verification and, more broadly, it must not be done in a way contrary to EU procurement principles of transparency, equal treatment, open competition and sound procedural management.

Understanding when social value is linked to the subject matter of the contract

In order to be able to evaluate social value criteria at tender stage under the Public Contracts Regulations, they must be linked to the subject matter of the contract.[36] This can be challenging to achieve in practice and, as a result, may lead some procurement officers to leave out social value requirements from their tender processes for fear of non-compliance. Consequently, it is only by getting to grips with this concept that social value opportunities in procurement can be maximised. There are limits to the types of requirement that may be considered to have a close enough relationship to the contract's subject matter in order for it to be legitimate to assess as part of tender evaluation under the Public Contracts Regulations.

Considering how social value forms part of the core of what is being procured will help ensure that social value is linked to the subject matter of the contract. Where social value is achieved at the core of a contract, a much greater impact can be achieved than through social value add-ons that do not strike at the heart of what the employer is seeking to achieve through the relevant contract. Where the social value requirements in a contract are closely aligned with what is at the heart of an organisation's overarching social value policy it is easier to demonstrate that the social value requirement is a critical element of what the employer is seeking to achieve through that contract. This can be demonstrated where it strikes to the heart of an employer's social value policy. However, regardless of the overarching policy position, individual procurement objectives will differ on a contract-by-contract basis. The key is to spend time at pre-procurement stage to identify those objectives that are core or critical to successful contract delivery and to record those decisions, reflecting them in the procurement documents, award criteria and the underlying contract.

Examples of where social value can be directly connected to the subject matter include:

- targeted recruitment and training for construction related skills in a large housing development;
- a contract to refurbish a school under which the contractor engages with students about careers in the construction sector;
- effective waste management and environmental management systems on site; and
- infrastructure contracts that look to promote opportunities to SMEs in the local area.

Where social value outcomes are identified that do not strike at the heart of the contract, employers can still build them into delivery as a contract condition. The difference is that where they are not linked to the subject matter of the contract they cannot be evaluated at tender stage.

Applying social value in practice

The need for a social value policy

What is core to an employer's overarching corporate policy is core to what it purchases. A social value policy will not in itself ensure social value is linked to the subject matter of every contract let by that employer, but can facilitate consideration of whether it is, pre-procurement.

A social value policy includes setting the social value objectives for an organisation and agreeing how and when the organisation will seek to implement them. Once established, employers can use this policy to take account of their social value core priorities before setting social objectives for each procurement and how they can be delivered through the contract terms.

Setting social value objectives for each contract/commissioning process is a critical part of the procurement of social value because:

- good procurement practice determines that tender/contract requirements should be measurable and capable of being monitored and verified;
- adopting a rational process for setting targets is important in obtaining the support of the whole procurement and contract management team;
- the objectives should be compatible with other contract priorities like quality, timely delivery and affordability; and
- it helps the employer to deliver against agreed objectives.

Good practice is to set social value objectives for each procurement and then monitor them throughout the procurement process, ensuring a clear and consistent approach.

A policy also enables employers to achieve a consistent approach to social value. An organisation's social value core priorities should be linked to its core activities. For example, an organisation that procures infrastructure maintenance works is likely to find that social value linked to training and employment in that sector is relevant to their organisation, as it will support the delivery of their principal ambitions.

The following are some examples of social value objectives that employers may wish to focus on in their social value policy:

Payment of a living wage

Many organisations that engage with social value are keen to require contractors to pay their employees a Living Wage (as distinct from the National Living Wage that is a legal requirement (see Chapter 4 for a discussion of the non-legal issues). However, there are legal limits to what can be done through the procurement process to impose a wage structure that goes beyond national minimum wage.

The Scottish Government has taken an innovative approach to facilitating consideration of whether contractors are Living Wage employers when evaluating bids. In the Scottish Government Statutory Guidance on the 'Selection of Tenderers and Award of Contracts'[37] the Scottish Government explains that it is possible to consider whether Living Wage is paid to employees as part of the evaluation of fair working practices of bidders. This approach demonstrates that it is still possible to set fair working practices as a policy objective, albeit that care must be taken when building it into underlying procurements so as to not fall foul of procurement law.

Targeted recruitment and training/tackling worklessness

In the construction sector there are fairly well-established processes for setting targets for 'new entrants' to the industry such as:

- a calculation of the overall labour requirement to deliver the contract e.g. number of people or person weeks; and

- a judgement of what percentage of these can reasonably be delivered by apprentices and new trainees, in the context of other contract requirements like quality, cost and timely completion.

Consequently policy objectives in relation to tackling recruitment, training and worklessness can be effectively achieved through contracts. The key will be to ensure policies are not too prescriptive so as to give procurement officers flexibility to meet the objectives in a way that is achievable when procuring each individual contract.

Tackling worklessness strikes to the core of a key social value outcome of tackling poverty itself. Some excellent examples of how this has been achieved in practice through public procurement is available through a report by Richard Macfarlane (2014) with Anthony Collins Solicitors LLP for the Joseph Rowntree Foundation entitled *'Tackling Poverty through Public Procurement'*, which makes for critical reading for any employer seeking to achieve the same.[38]

Impact on the local economy

It is often popular for employers to aim to procure locally in order to benefit the local economy. However, it is not lawful under the EU freedom of movement rules to restrict supply to that which is local as this is anti-competitive. That said, policy objectives can be set to ensure opportunities are made available to local bidders, which in practice can be achieved at procurement stage by notifying local organisations about the opportunity to tender for contracts/subcontracts. Whether opportunities are ultimately awarded to those local organisations will depend on their suitability, capacity and price, and in relation to subcontractors, this is a decision for the contractor.

Support to SMEs and access for SMEs

The Local Government Association construction strategy of 2018 highlights that 99 percent of construction firms are SMEs, and over 40 percent of jobs in construction are self-employed.[39] This demonstrates that where employers want to achieve social value through works projects they must be robust in what they require contractors to deliver down the supply chain.

A policy objective can be set to ensure that a level playing field is created that enables SMEs to compete effectively for tenders. This can be achieved by:

- drawing the attention of SMEs to contract opportunities. Provided the advertisement requirements under the Public Contracts Regulations and UK law are complied with, it is possible to advertise the opportunity in other locations to help secure social value; or
- requiring the main contractor to open up their own supply chain.

In addition, where contracts have a value exceeding the relevant EU tendering threshold under the Public Contracts Regulations, in the pre-procurement stage employers must consider whether division of the contract into smaller lots would be appropriate. Where the contract value is below threshold it is still best practice to consider doing so as smaller lots can open the contract up to a wider range of bidders, particularly SMEs. Adopting this approach is much more likely to achieve social value. Best practice is to consider this widely, but in particular look at whether it is appropriate to divide the contract into lots.

A policy that seeks to provide greater opportunities to SMEs through advertisement and/or smaller contracts can effectively be developed at procurement stage to meet the specific needs of the contract being procured.

Support to social enterprises and the 'third sector' and access for those organisations

The same approach to SMEs can be taken in order to ensure social enterprises and 'third sector' organisations have equal access to procurements. In addition, Regulation 20 and Regulation 77 of the Public Contracts Regulations provide that contracts can be reserved for sheltered workshops, sheltered employment programmes and specified service contracts, provided certain requirements are met. However, the nature of organisations that contracts can be reserved for is limited and less likely to be applicable to the construction sector. Consequently, whilst it may be possible and appropriate to set a policy objective to maximise opportunities for third sector/social enterprise organisations, it is less likely that a prescriptive policy to award reserved contracts where possible will be achievable when procuring works contracts (although it may still be relevant at policy level in relation to procurement of suppliers and services).

Social value impact through the supply chain

A policy seeking social value through the supply chain can increase the impact achievable through individual procurements. Successful delivery of social value down the supply chain is directly linked to the social value requirements that are placed on the main contractor.

Environmental sustainability

A policy objective linked to environmental sustainability can readily be delivered through works contracts. For example, environmental sustainability requirements can be achieved from a contract through the following mechanisms:

- Sustainability Action Plan: develop a Sustainability Action Plan that addresses environmental management in the delivery of works and demonstrates a commitment to using working methods, equipment and materials that will improve the sustainability of the contract requirements.

- Environmental Management System: put in place or require an Environmental Management System that applies specifically to the works for the duration of the contract.
- Imposing obligations on third parties: when contractors purchase supplies, equipment or services from third party contractors, contractors can be asked to require that third party contract to meet the requirements of the Sustainable Action Plan above.
- KPIs: require contractors to adhere to environmental key performance indicators. Consider how those KPIs are to be effective (for example, prolonged failure linked to termination rights or achievement linked to incentives).

Preliminary market engagement

A key solution to the tension between the need to prescribe what is being procured and the need to facilitate innovation is effective use of preliminary market engagement. Engaging with the market pre-procurement means employers can understand better the opportunities available to them.

In order to get the most out of public spend it is important for employers to clearly specify what they are seeking to procure in their invitation to tender (ITT). A thorough understanding of what the market has to offer in advance of going out to tender ensures that not only is maximum social value secured from the contract, but that consideration is given to how the social value is linked to the core of what is being procured. This enables specific social value requirements to be set at procurement stage and in contract terms that are quantifiable, reasonable and deliverable.

As with all elements of the procurement process, it is important to ensure that proposals for pre-procurement market engagement relating to social value will not violate the EU Treaty principles of transparency, equal treatment and proportionality.

Market engagement activities that can be undertaken in a way that ensures EU Treaty principles are upheld include:

- 'Meet the buyer' events – either directed towards one particular contract, or more generally focused on how the employer commissions its works. These can also be organised by umbrella bodies (such as the local Chamber of Commerce) and so enable contractors to meet a number of potential buyers at one event.
- Soft market testing – seek to investigate how a market feels about a potential contract, new ideas and potential choices that the employer could make.
- 'How to' guidance for bidders – employers can produce guides for bidders on how to do business with them. These need not be contract-specific and can provide information on the employer's arrangements for purchasing, including (for example) general information on the organisation's tendering procedures.

It is very important to keep records of what took place during any pre-market engagement, not least because it ensures information arising from it must be shared with all bidders at the procurement stage to ensure competition is not distorted. Records will also provide evidence of consultations undertaken as a consequence of the Employer's duties under the Social Value Act.

Social value at pre-procurement stage

The obligations of the Social Value Act 2012 must be complied with during the pre-procurement stage. As a minimum, each of the social value objectives set out at policy level should be considered, but it may be possible that the nature of the contract allows social value to be achieved that goes beyond that prescribed in such a policy.

At this stage, employers must consider and set specific social value objectives for the contract being let. Once set, it is sensible to consider and plan for how this can be achieved through the procurement process to ensure success.

Preparing the procurement documents

Procurement documents must be freely available to bidders when publishing the OJEU contract notice or the invitation to confirm interest. The Public Contracts Regulations define 'procurement documents' very broadly and include documents such as the OJEU contract notice and other notices, the technical specifications, the proposed contract conditions and any descriptive document, as well as any 'additional documents' relating to the tender process. It is important to stop and consider what procurement documents must be drafted at the outset of any procurement and to consider how to build social value requirements into those documents in a way that is linked to the subject matter of the contract.

Contract conditions and specification

Social value obligations should be built into contracts as contract conditions and/or items in the specification to ensure they are enforceable. Where the social value outcome is to be included as a contract condition and/or in the specification and is linked to the subject matter of the contract, bidders' ability to meet that outcome can be considered as part of the tender evaluation.

Invitation to tender

Good practice is to use the invitation to tender (or equivalent) to:

1. explain the social value requirements of the contract to bidders; and
2. assure bidders they are not being required to provide a disproportionate response.

Mentioning social value in contract advertisements

For contracts that are advertised via OJEU, employers must refer to the use of social considerations in the OJEU contract notice, prior information notice or other notice published as well as in the Contracts Finder advertisement. This need not be a detailed breakdown, but setting social value outcomes at pre-procurement stage enables a summary of these to be easily drafted into such notices.

Given this, at pre-procurement stage it will be important to consider and ensure that social value outcomes set are measured so that award criteria can be drafted in the tender documents that are sufficiently transparent and quantifiable.

Evaluating bidders' responses to social value requirements

Under the Public Contracts Regulations, a contract must be awarded on the basis of the most economically advantageous tender (MEAT), which means an evaluation of the price or an evaluation of the best price-quality ratio.

Social value can be considered as part of the quality assessment of a bid. An evaluation of the life-cycle cost of a contract when evaluating 'price' can also help ensure the sustainability of a contract.

Quality

Best practice is to include the award criteria and the weightings attached to social requirements in the ITT as quality aspects of the award criteria.

Employers should consider at pre-procurement stage how much weighting to give social value requirements overall and/or individually on a contract-by-contract basis to ensure the weighting given is proportionate considering all of the specific circumstances of that procurement. This will enable sufficient and considered drafting of the tender document.

Price

Employers can consider evaluating the life-cycle cost of a contract to ensure the sustainability of a contract. The following are some examples of key elements of life-cycle costing:

- costs of acquisition;
- costs of use (e.g. consumption of energy and other resources, maintenance costs, end-of-life costs such as collection and recycling); and
- costs relating to environmental externalities linked to the works during its life cycle as long as the monetary value of the cost is capable of being determined and verified. Examples include cost of emissions of greenhouse gases and other pollutant emissions and other climate change mitigation costs.

When evaluating the life-cycle cost an evaluation method must:

- be based on criteria that are objectively verifiable and that are non-discriminatory and does not unduly favour or disadvantage certain economic operators;
- be accessible to all bidders; and
- only require data from bidders that can be provided with reasonable effort by normally diligent economic operators including those from third countries party to the Agreement on Government Procurement or other international agreements by which the EU is bound.[40]

When evaluating life-cycle costing the following should be indicated in the procurement documents:

- the data to be provided by bidders; and
- the method which will be used to determine the life-cycle costs on the basis of those data.

Social value during the procurement process

How the tender process will be structured is decided pre-procurement. This should be adhered to during the tender process and the documents drafted pre-procurement should be relied upon.

Where procurement has been appropriately planned, documents properly drafted and procedures adhered to, social value requirements being evaluated before contract award will have been properly incorporated into the procurement documents. Once tenders are received, social value responses should be considered by properly applying the evaluation methodology, selection criteria and award criteria set out in the invitation to tender.

Social value during the contract term

All of the work that goes into achieving social value at the pre-procurement stage can be lost if effective contract management is not used to enforce the contract documents and ensure contractors comply with their social value obligations.

The key challenges to social value are interrelated. In particular, an under-appreciation of what social value can achieve can lead to ineffectual contract management as it contributes to a culture where social value is viewed as unimportant. However, where employers have gone to the effort of securing social value, and have awarded a contract to the successful bidder by taking social value into account, it is a huge waste of resources and opportunity to then not follow through with effective contract management. This is particularly true in the case of infrastructure projects where the total spend will most often be in the millions of pounds or in the case of substantial projects such as HS2 in the billions.

A further issue with ineffective contract management is that, if an employer fails to require successful bidders to perform key elements of the contract, the decision to award that contract could be subject to challenge on the basis that it breaches the principles of transparency and equality.[41]

The key to securing social value during contract delivery is contract management. Contractors will deliver on the elements of a contract they feel their employer wants, and the elements the customer is prepared to pay for. Therefore, proper and thorough contract management, coupled with an appreciation of the value of social value, can be crucial in achieving the benefits envisaged during the procurement process.

This means that Contract Managers should robustly handle non-performance by contractors, including social value requirements. It is good practice to give consideration in advance to how to respond where a contractor:

- fails to meet its contractual reporting obligations; and/or
- in its reporting has failed to demonstrate that they have fully implemented an agreed social value requirement or objective.

Where non-performance of social value outcomes arises, the most effective means of enforcement is to adhere to the terms of the contract by withholding partial or total payment until the omission is rectified (or otherwise as the contract sets out). Termination under the terms of the contract is generally an option of last resort. A preferable solution is to encourage the contractor to rectify the situation themselves. Ultimately this will save costs for both the employer and the contractor.

Effective monitoring of a contract should assist Contract Managers in avoiding the 'nuclear option' of termination.

Monitoring contract performance, for example through key performance indicators, is not only essential for ensuring social value targets are met, but also provide the following additional benefits by giving insight into:

- the extent to which the employer is achieving social value through their works contract;
- the proportion of social value added in relation to additional procurement costs incurred;
- what contractors are willing and able to provide;
- areas in which trends of non-performance are arising; and
- how social value can be delivered by contractors, leading to clearer contract award criteria in future procurements.

The technology of social value monitoring on site should also be considered. With sophisticated site management and modelling software packages being used on site, how can these be used to monitor, and perhaps model, social outcomes during delivery?

At present, it is likely that they will only be used to deliver the most high-level data, but if social value is best achieved when it is at the core of a contract's requirements then surely it should be more imbedded within the site management tools already used to manage and monitor other key works deliverables.

Social value post contract

After a contract is completed, best practice is for an organisation to review that contract's social value impact against its social value policy and goals. The social value data, which has been gathered during the contract, provides a wealth of information that can be analysed to understand areas where social value has been achieved as anticipated and areas that have not worked quite so well. This information is invaluable for informing future procurements as employers can tailor social value requirements to maximise impact in areas that have proven to be effective in the past.

Many organisations undertake an annual review of their social value policy, which is also good practice. Understanding the social value impact of past contracts feeds into this annual review as it informs amendments and improvements to the policy. Ongoing learning is key to maximising social value impacts.

Summary

Social value presents huge potential for employers to maximise public spend and increase value for money in their construction projects. The commitment to this in post-Brexit government and local government policy publications focusing on construction and infrastructure projects demonstrates that a social value objective to public sector contracting will continue even when the UK leaves the EU. With a renewed focus on 'localism', it may be that in the long term there is even greater focus on this area of public contract tendering.

As this chapter demonstrates, in order to fully achieve the potential of social value, employers must consider social value at all stages, from policy setting to pre-procurement, and through to procurement and contract award. However, even where procurement processes focus on social value, there is a risk that social value will not be enforced by contract managers where it is not fully embedded into the core of the contract, or there is a failure to recognise the significance of social value. Bringing social value into the core of the contract at the procurement stage is a key way to address this issue, but it also requires active involvement from contract managers who appreciate the significant role social value can play in driving value for money in public contracts. Effective pre-market engagement also has a role to play in demonstrating the social value that bidders can offer and is promoted by public policy, facilitating an approach that secures maximum social value without inhibiting innovation.

In addition, a thorough understanding of the legal framework that surrounds social value is critical to overcome the challenges that employers face when seeking to procure social value. It is only by understanding the parameters of what the law allows that the opportunities social value presents can be harnessed.

Where this can be achieved at each stage of the procurement process, the opportunities for delivering social value through the procurement and delivery of works and infrastructure projects is vast. This is exciting for public sector employers because it gives them the opportunity to use procurement as one of the solutions to improving lives, communities and the environment within their delegated authority.

Key points

- The current focus of the UK government on infrastructure spend represents significant opportunities to deliver social value through such projects.
- The Social Value Act requires those procuring public services contracts to consider how to achieve social value at pre-procurement stage.
- The Social Value Act is deliberately non-prescriptive – it encourages contracting authorities to capitalise on the opportunity that social value presents.
- Applying the principles of the Social Value Act across all public contracts, whether or not for services, is best practice.
- Consideration of social value objectives at pre-procurement stage ensures that social value delivery is maximised.
- When procuring social value through public contracts, the procurement rules set out in the Public Contracts Regulations 2015 must be followed.
- Understanding the public procurement legal framework is the key to successfully achieving social value through procurement of public contracts without legal challenge.
- When procuring social value, it is critical that the social value requirements of a contract form part of the award criteria. In order to do so, under the Public Contracts Regulations 2015, those social value requirements must be sufficiently linked to the subject matter of the contract.
- In order to ensure that social value is delivered through such a contract, the requirements must form part of the contract terms awarded and proactive contract management used to enforce those terms.

Disclaimer: Whilst every effort has been made to ensure the accuracy of these materials, advice should be taken before action is implemented or refrained from in specific cases. No responsibility can be accepted for action taken or refrained from solely by reference to the contents of these materials. © Anthony Collins Solicitors LLP 2018.

Notes

1 Section 3(1), European Union (Withdrawal) Bill 2017–2019.
2 Procurement Policy Note 09/16 – Procuring for Growth Balanced Scorecard.
3 HM Government, 'Building Our Industrial Strategy', Green Paper, January 2017.
4 HM Government, 'Industrial Strategy Building a Britain Fit for the Future', November 2017.
5 Local Government Association, 'National Construction Category Strategy', 2018 Edition.
6 Procurement Policy Note 01/18 – Supply Chain Visibility.
7 Section 1(3), Social Value Act.

8 Section 1(7), Social Value Act.

9 Section 1(1), Social Value Act.

10 Section 1(8), Social Value Act.

11 See Section 54(2(b) of the Modern Slavery Act 2015, and Regulation 2 of the Modern Slavery Act 2015 (Transparency in Supply Chains) Regulations 2015.

12 The general duty is set out at Section 3 of the Local Government Act 1999.

13 Department for Communities and Local Government 'Revised Best Value Statutory Guidance', March 2015.

14 Equality Act 2010, Section 149.

15 Section 1(11)(e), Social Value Act.

16 'Community Benefits Delivering Maximum Value for the Welsh Pound': http://gov. wales/docs/dpsp/publications/valuewales/140904-community-benefit-report-en.pdf.

17 n15 pages 10–11.

18 Procurement Reform (Scotland) Act 2014, section 9.

19 n17 section 15, and see also section 24 for a definition of 'community benefit requirement'.

20 n17, section 25.

21 Article 70, Directive 2014/24/EU.

22 Recital 97, Directive 2014/24/EU.

23 Recital 78, Directive 2014/24/EU.

24 Article 67(2), Directive 2014/24/EU.

25 Articles 18(2) and 56(1), Directive 2014/24/EU.

26 *European Commission v Netherlands* C-368/1[2013] All ER (EC) 804 ('the *Dutch Coffee Case*').

27 Article 43, Directive 2014/24/EU.

28 Recital 93 and Article 67(2), Directive 2014/24/EU.

29 Recital 97, Directive 2014/24/EU.

30 Regulation 67(2) Public Contracts Regulations 2015.

31 Recital 1, the Consolidated Directive.

32 *Laboratori Bruneau Srl v Unità Sanitaria Locale RM/24 De Monterotondo* [1994] 1 CMLR 707 ('*Laboratori Bruneau*').

33 *Concordia Bus Finland Oy Ab (formerly Stagecoach Finland Oy Ab) v (1) Helsingen Kaupunki (2) HKL-Bussiliikenne* (C513/99) [2003] 3 CMLR 20 ('*Finnish Buses*').

34 *EVN AG and Another v Austria (Stadtwerke Klagenfurt AG and Another, intervening)* (C448/01) [2004] 1 CMLR 22 ('*EVN*').

35 *European Commission v Netherlands* C-368/1[2013] All ER (EC) 804 ('the *Dutch Coffee Case*').

36 Public Contracts Regulations 2015, Regulation 67(2).

37 The guidance can be accessed at: www.gov.scot/Resource/0048/00486741.pdf.

38 Richard Macfarlane with Anthony Collins Solicitors LLP for the Joseph Rowntree Foundation, 'Tackling Poverty through Public Procurement', April 2014 which is available to download from www.jrf.org.uk/report/tackling-poverty-through-public-procurement.

39 Local Government Association, 'National Construction Category Strategy', 2018 edition, page 14.

40 Regulation 68, Public Contracts Regulations 2015.

41 The Public Contracts Regulations 2015 Regulation 18; as discussed in The Scottish Government's 'Community Benefits in Public Procurement' at page 28.

3 Theoretical justification for social value

Andrew Knight

This chapter develops a theoretical foundation and justification for social value from moral philosophy. We draw on specific moral theories from utilitarian, Kantian and existential philosophers to explore why promoting social value is morally praiseworthy. However, we will also demonstrate some of the limitations of these theories by illustrating both contradictions between theories and the counterintuitive results that sometimes arise. The general conclusion is that social value promotion has strong theoretical support, however, we end on an existential note, acknowledging that although moral theory has the power to inform our decision making, ultimately we are condemned to be free as autonomous agents.

Some conceptual distinctions between moral and non-moral reasons

To question why we morally *ought* to maximise social value, it is necessary to conceptually distinguish non-moral reasons from moral reasons for acting, so the grouping of reasons do not get confused. There are many non-moral reasons for enhancing social value, for example, to comply with the law, as in the case of social value acts of parliament. The same can be said if a client requires that a submission for tender must include a provision for social value. Compliance by a contractor is therefore required for contractual and not moral reasons. Finally, many companies have strong ethical statements around broad issues such as 'making the world a better place', however, if this is mere prudential reasoning, i.e. the board members of the firm believe that in the long term they will maximise profits by following what appears to be an ethical objective, again this is not a moral reason, in and of itself, although it may *accord* with a moral reason.

In Kant's 1785 *Groundwork of the Metaphysic of Morals,* he is keen to distinguish moral reasons from reasons which merely accord with our moral duty. He cites the case of a shopkeeper in the days before individual items in a shop were labelled with a price (Kant and Paton, 1991: 63). Kant argues that even though there is a duty not to charge an inexperienced customer more than anyone else, the decision not to overcharge may also be motivated from natural self-interest, since the shopkeeper would not want his reputation damaged; for example, if word spread that he was charging different customers different prices. In short, according to Kant, actions which merely *accord* with our moral duty, for example to act honestly to preserve reputation, are not morally praiseworthy, whereas actions performed *from* moral duty are morally praiseworthy. Doing the right thing because it is the right thing is distinctly different to performing the same action, with the same consequences, but for other motivations or inclinations. It is useful in philosophical analysis to start with some conceptual tidying and we will return to Kant later, but we now consider a very influential group of thinkers commonly grouped together as utilitarians. Their approach is quite different to that of Kant's, since it is the consequences of actions that morally matter to the utilitarians, rather than the intentions of those actions.

Utilitarianism

Utilitarianism, as the name suggests, is concerned with maximising utility or happiness. As a moral philosophy, modern utilitarianism is related to concepts of social value, because it was developed by Bentham and others as part of a social agenda to improve society by bringing rational, evidence informed policy and laws to England during the eighteenth century. Hence, utilitarianism is practical as well as theoretical.

So according to Bentham and Harrison (1967: 125), 'Nature has placed mankind over the governance of two sovereign masters, pain and pleasure. It is for them alone to point out what we ought to do, as well as to determine what we shall do'. So for example, if redistributing wealth from the top ten percent of income earners to the bottom ten percent increases the pleasure of the poor more than the pain to the rich, i.e. overall there is a net increase of pleasure in the population, then this redistributive exercise is morally praiseworthy. This example captures the importance of equality in utilitarian calculations. In essence, for the utilitarian, everyone counts, rich or poor, and hence the theory and practice plays a central role today in government policy and law making, with social value policy having a particular 'for the greater good' utilitarian flavour.

An important criticism of utilitarianism is the assumption that pleasure is equated with what is morally good. As Russell (2015: 702) points out 'J.S. Mill, in his utilitarianism, offers an argument that is so fallacious, that it is hard to understand how he can have thought it valid. He says: Pleasure is the only thing desired; therefore pleasure is the only thing desirable'. There are many examples of things which may be pleasurable, but that hardly seem desirable from a moral perspective, for example, the case of consuming highly addictive street drugs. Additionally, imagine the case of a new motorway built to aid economic development in a city, whilst there may be a net increase in happiness across the whole population, if many traditional homes are demolished and families displaced by the construction process, the project still feels morally problematic (for further discussion and important critics of Utilitarianism see for example Moore 1959; Smart and Williams 1973).

Interest in practical ethics started to emerge again by the 1970s and 1980s. Peter Singer (b. 1946), for example, has been key to a regeneration of awareness in applied ethics in a variety of contexts (see Singer 2008). However, the old utilitarianism has evolved into what is now broadly termed consequentialism, this shares the focus on consequences with the original utilitarian theories but removes the assumption of hedonism, i.e. things other than pleasure can be the optimisation target, such as freedom or human dignity.

Although consequentialism still has the problem of how to measure the impact of actions, its evolution has meant its continuing popularity as a moral justification for a range of social, political and legal policies. Hence, the spirit of Bentham still provides the moral, and sometimes radical, imperative for a social value agenda.

Kantian ethics

Rather than focusing on the consequences of actions to evaluate moral worth, Kantian ethics focus on the intention of the person carrying out the act to evaluate moral praiseworthiness: 'It is impossible to conceive anything at all in the world, or even out of it, which can be taken as good without qualification, except a *goodwill.*' (1991: 59). The goodwill is a central concept in Kant's moral system, because to act from a goodwill is what defines an act as morally right. From an alternative viewpoint, Aristotle's virtue-based ethics focuses on the character traits of the actor, classical virtues include traits such as courage, temperance and patience (2009). However, these virtues are only conditionally good according to Kant. He reinforces his point by stating that even when circumstances stifle our ability to act, a good will would still 'shine like a jewel for its own sake as something which has its full value in itself' (1991: 60), thus reinforcing his concern with intentions rather than consequences.

Kant's writing is generally considered challenging. However, the core principles which underlie it are clear. Warburton (1998: 177) states 'Our moral duty arises from our respect for the moral law. The moral law is determined by what Kant calls the categorical imperative.' So when we act with a good will, we are acting from our duty. In the *Groundwork,* Kant argues that when acts are undertaken that are in accordance with duty, they are not morally praiseworthy since the motivation could result from our inclinations. This difference between the same action being in *accordance* with duty versus *from* duty, is critical from Kant's perspective, as stated at the start of the chapter. The first motivation is not worthy of praise: the second is. However, this can lead to some rather counter intuitive situations. Imagine two workers, Ms A. and Mr B., are employed on a social value enterprise to help young people gain employment in construction. Ms A. loves her job and gets a real warm glow from the pride she has in seeing the young people gain employment. Mr B. really dislikes spending time with the young people, but works to his best ability to help them gain employment. From a Kantian perspective, if Ms A. is doing the work because she enjoys it and it makes her feel good then this is not morally praiseworthy, since it is an action merely in accordance with duty. Whereas, Mr B. is morally praiseworthy since he is working because it is his duty. It could be argued that a weakness of Kant's approach is that it evaluates the importance of human reason to the exclusion of emotions. However, an ethic focused on reason and recognising the moral law may provide to be more applicable in organisational environments in the context of social value decisions, since rational discussion is a typical mode of group decision making.

A further question to pose is: what is the origin of the moral law? Kant's does not tell us what the moral law is, his ingenious solution to the problem is to make us all, rational autonomous persons, creators of the moral law, using the categorical imperative, which appears in various formulations throughout the *Groundwork*. The categorical imperative is logically unconditional and can be contrasted with a hypothetical imperative. Thus, 'do not inflate claims of social value in tender documents' is categorical whereas, 'if I want to win the tender for a new project, then I should inflate social value claims' takes the IF/THEN logical form and is clearly conditional. Here we will only focus on two formulations of the categorical imperative to demonstrate how Kant's duty-based ethics may provide a driver for social value.

The first formulation of the categorical imperative can be summarised as 'I should only act on rules, where I rationally want others to also act on those same rules'. So for example, if in a building project I accept my surveyors falsifying day work costing sheets to make more profit on variations to the final account, is this morally acceptable? Using the universalisability formulation of the categorical imperative: would I rationally want other parties, for example subcontractors, also falsifying records in their dealing with me? In short no, since trust between the parties would break down. Therefore, I ought to act on the maxim 'be honest and transparent in my business dealings' because if I then accept falsifying records, I'm making an exception of myself. Of course, if not caught out, this sharp business practice might very well be beneficial to my organisation in terms of profit and indeed my personal profit-related bonus, but to act in a manner in which we would not want others to act, is clearly not showing respect for the moral law. Therefore, the first formulation of the categorical imperative, in this example, supports what can be described as professionalism.

A second formulation of Kant's categorical imperative states, 'Act in such a way that you always treat humanity, whether in your own person or in the person of any other, never simply as a means, but always at the same time an end' (1991: 91). Although there is debate as to whether this is a genuine reformulation of the same imperative, we will accept nevertheless, the principal that treating other people as rational autonomous beings, like ourselves, involves treating others with respect. So for example, the growing problem of modern day slavery can clearly be seen to violate both the above formulations of the categorical imperative. First, because I can't will others, when they need cheap labour, to own slaves since this can't be rationally universalised, i.e. I would not want other slave masters owning me as a slave. On the second formulation, slavery treats other human beings as mere property in the most degrading manner. Slaves have no autonomy to plan their lives and make autonomous choices. This situation is one of the clearest examples of treating persons as a mere means, rather than an end in themselves. But what about more subtle problems where the decision making is more complex and the morality less immediately obvious?

Decisions between alternatives

In Chapter 10, a case study of the Dounreay nuclear power station is considered. From the initial development in the 1950s, there are obvious but sometimes unequal benefits brought to the area. The 'Atomics' were the new, highly educated workers who moved to Dounreay and their impact on the local original residents was substantial. This was in terms of economic impact, public service improvements and the cultural environment. However, there were significant costs and many of the original residents did not benefit to the same degree as others. Nevertheless, if a consequentialist approach to the decision making had been applied, at least in principal, a conclusion could have been calculated that the new power plant was for the greater good. Accepting the practical difficulties with measuring costs and benefits, a similar consequentialist approach could provide a result for future developments of the site after decommissioning. But what about the application of Kantian reasoning to such a large-scale project? And how do we reliably evaluate alternative uses for the decommissioned site?

We could pose questions as to whether the original residents were merely a means to an end, for example the aim for power production for the rest of the country, rather than being treated as ends in themselves by the originators of the project. The answers to these questions, over half a century later, become speculative. However, it is reasonable to suggest that more recent policy formulation regarding the future of the decommissioned site has a Kantian flavour. For example, the remaining residents appear to be trusted as partners in the co-creation of social value, thus respecting them as autonomous individuals worthy of respect. The acknowledgement by the authorities that solutions for the site consists of 'alternative' rather than 'best' options, additionally seems to indicate an approach that is more Kantian. So, although a Kantian approach may not provide a clear summation of the 'right option', its strength in real world complex situations is that it does not try to impose a 'best' solution; rather, those involved in decision-making recognise their duties to others, like themselves, as people worthy of respect with views that ought to be fully considered in any decision making.

The two normative ethical theories explored above, Consequential and Kantian, both add significant theoretical weight to the moral case for improving the world in terms of the broader social value agenda beyond baseline profits and financial cost minimisation. However, we have raised the question: how do we decide between competing social value projects?

Utilitarianism has been traditionally associated with the objectification and measurement of costs and benefits. Bentham's original work provided a comprehensive approach to considering a range of factors involved in pleasure or pain including: intensity, duration, certainty, propinquity or remoteness, fecundity and purity (Bentham and Harrison 1967). In many ways, this is one of the ancestors of modern cost benefit analysis and today throughout social value policy, measurement of impact is a central tenant. But just how in practice does one validly and reliably value the enjoyment of a quiet park, or the pleasure of attending the community centre Bingo game, or even a Chopin concert? Bentham clearly believed that going through his calculus of happiness, the *summum bonum* could be calculated. However, developments in the utilitarian school of thought were already casting doubt on this reductive approach. For example, Mill (Mill and Crisp 2010), in response to criticism that utilitarianism was a 'swine morality' for promoting base pleasures, started to defend a non-commensurable system of higher and lower pleasures, based on whether the higher or lower faculties were exercised. Unsurprisingly, philosophy was a higher pleasure and only those who had experienced these higher pleasures were fit to judge! Of course, some may argue this is mere cultural elitism.

From a duty-based Kantian ethic, social value impact measurement is even more problematic, since it is the maxims which motivate the actions, rather than the results that matter, measured or not. Core to the Kantian attitude is recognition of our duties to others and respect for people who ought to have their voices included in the decision-making process as discussed in the above case study.

In summary, although the above ethical theory provides a clear reason why we ought to maximise social value, it might provide a less clear validation of current methods used to measure impact. This in itself is interesting since it may result in more understanding and questioning, not of the good of social value, but on the reliance placed on measurement in decision making and evaluation. To develop these ideas further, the final section of this chapter draws on some of the more radical philosophy of the twentieth century, namely existentialism.

Some existential conclusions

Sartre explains why traditional moral theories, such as those discussed above, and religious guidance fail to clearly show us what we morally ought to do. His most vivid illustration is the powerful vignette of the student who approaches him for moral guidance.

The student finds himself pulled between two solutions to a moral dilemma. The case is set in World War II France, where the student's mother is in a state of despair. Her husband is a collaborator with the occupying forces and her other son has been killed in the 1940 offensive, all she now has is her remaining son. The student goes to Sartre to ask for advice, he is considering two possibilities: stay at home and take care of his mother or go and help fight with the Free French Forces. Sartre uses the example to demonstrate the impossibility of evaluating the morality of the two options. From a broadly consequential position, he argues that it is impossible to compare the very direct impact on staying and caring for his mother against the far greater, but indeterminate, impact on joining the war effort, which he described as 'an ambiguous action which might vanish like water into sand and serve no purpose' (1989: 35). Sartre also argues that Kantian ethics provide no help for the young man since 'if I remain with my mother I shall be regarding her as the end and not the mean: but by the same token I am in danger of treating as a means those who are fighting on my behalf, and the converse is true…' (1989: 36).

Sartre then extends his analysis of the case to demonstrate why our moral feelings do not even help because we cannot know what we feel until we have performed the relevant actions. Additionally, taking counsel from others is no solution, since we can choose our advisor to fit our preferred outcomes, for example, we can seek the advice of either collaborator or resistance priests. All this results in the conclusion that ultimately the student *himself* must make the decision to fight or stay with his mother; it is his choice, no one else, and no ethic can make the decision on his behalf. In Sartre's words 'you are free, therefore choose – that is to say, invent. No rule of general morality can show you what you ought to do: no signs are vouchsafed in this world' (1989: 39). This is the radical freedom at the centre of existentialist thought. We have argued that moral theory does *generally* provide a solid foundation for the moral case for social value, but we conclude that at the margins between different choices of project, or on the issue of impact evaluation, ultimately, we must recognise the limitations of measurement in a world overly concerned with objectifying decisions. To precis Sartre, we must make our own judgments: we are condemned to be free: we are responsible for our own choices.

Before closing this chapter, we will briefly mention a further existential category, which may help justify the social value agenda, that of *authenticity*. Existentialist thinking centres on human freedom (Warnock 1970: 1). The term 'existential' simply means that existence precedes essence. As a central case, in Sartre's work, humans have no predetermined essence, we are free to determine our own essence through our actions during our lives. He terms a human being as a, 'Being-for-itself' in contrast with the inanimate realm of objects, such as a book or paper-knife, which he terms a 'Being-in-itself'. In Sartre's example of a paper-knife, the essence of the object, its purpose, is determined before its manufacture by the designer (1989: 26). Contrary to the Being-for-itself, with the Being-in-itself, essence precedes existence.

According to Sartre, one of the greatest dangers for human beings is not recognising our own freedom to create our own essence through the actions we choose; in short, the danger of not living authentically. In this way, a human life can be seen not as predetermined, but as a work of art, created and developed throughout one's life. Sartre contrasts this with the person who fails to recognise this freedom and lives a life of 'bad faith'. He develops the vignette of the waiter, who rises every day at 5 am to work in the café, Sartre says, 'Let us consider this waiter in the cafe. His movement is quick and forward, a little too precise, a little too rapid. He comes toward the patrons with a step a little too quick. He bends forward a little too eagerly; his voice, his eyes express an interest a little too solicitous for the order of the customer' (2010: 82). The important point is not that the man is acting like a waiter, but that he is being defined by his role. He has trapped himself into a self-imprisoned life of bad faith.

Sartre (1989) is particularly critical of capitalism since he claims it appears to reduce our freedoms when we develop lifes around slavish patterns of work and consumption. In this state of inauthenticity, we do not acknowledge the truth, i.e. that we are free to walk away, to do something more fulfilling, more authentic. Although this may appear to be too radical for most of us, Sartre does acknowledge that there are things we cannot change; for example, our parents, our place and time of birth, and so on, he calls this our 'facticity'. Nevertheless, his bad faith point is well made. Despite our facticity, we have far more freedom to live more authentic, self-determined lives than most of us ever acknowledge: we can do things differently! This claim provides the existential motivation for the individual to reconsider a life locked into profit maximisation and consider the social value alternatives.

There sometimes appears to be a causal necessity that underpins the neoliberal consensus that the profit maximising market always knows best. It takes both drive and commitment to think things could be done differently and to make these things happen. Social value entrepreneurs who see things differently, who want to change the status quo and improve the world, embrace Sartre's call to live more authentically and do things differently. However, too often the task of improving genuine social value seems extremely challenging. Hence, small changes, such as the steps to incorporate social enterprises into parts of the supply chain (discussed in Chapter 4), become very important. To conclude, even if most individuals would struggle to live a more authentic life, to embrace the radical changes that this would cause both personally and professionally, we should admire and support the projects of those who do take a different route. Finally, even though we follow Sartre in believing ultimately we are all free to choose, we are also responsible for those choices, and just repeating the mainstream views and fitting within the orthodoxy, is also a choice, a choice not to change, and for this too, we are responsible.

Summary

This chapter commenced by drawing on Kant to distinguish between moral and non-moral reasons for acting. We then explored two areas of theory that broadly support a social value agenda: duty based ethics and consequentialism. Using Kant's categorical imperative, it was possible to argue along two lines that decisions which could be rationally universalised and decisions which respected other people as ends in themselves, could provide ethical direction in social value decision making. Furthermore, consequentialist and utilitarian thinking was also identified as moral theory to support a challenge to the orthodoxy of the neoliberal profit maximisation ideology of the free market. However, a caution was sounded in respect to choices between alternative proposals, where even the possibility of accuracy in decisions was questioned. Finally, a call to social value from an existential imperative reminds us of the fact that we are ultimately free to choose the type of life we wish to live.

Key points

- Moral and non-moral reasons for acting should be distinguished.
- Duty based ethics can provide support for a social value agenda.
- Consequentialism and utilitarianism typically also provide support for social value.
- Care should be taken when comparing alternative options in social value projects.
- According to the existentialism, we are condemned to be free and live the life we choose.

References

Bentham, J. and Harrison, W. (1967) *A Fragment On Government: And an Introduction to the Principles of Morals and Legislation*. Oxford: Basil Blackwell.

Kant, I. and Paton, H.J. (1991) *The Moral Law*. London: Routledge.

Mill, J.S. and Crisp, R. (2010) *Utilitarianism*. Oxford: Oxford University Press.

Moore, G.E. (1959) *Principia Ethica*. Cambridge: Cambridge University Press.

Russell, B. (2015) *History of Western Philosophy*. London: Routledge.

Sartre, J.-P. (1989) *Existentialism and Humanism*. London: Methuen.

Sartre, J.-P. (2010) *Being and Nothingness*. London: Routledge.

Singer, P. (2008) *Practical Ethics*. Cambridge: Cambridge University Press.

Smart, J.J.C. and Williams, B. (1973) *Utilitarianism: For and Against*. Cambridge: Cambridge University Press.

Warburton, N. (1998) *Philosophy*. London: Routledge.

Warnock, M. (1970) *Existentialism*. Oxford: Oxford University Press.

4 Creating social value within and between organisations

Ani Raiden, Martin Loosemore, Andrew King and Chris Gorse

This chapter discusses different ways of creating social value. We first present ideas about strategic, organisation level thinking. We highlight a partnership model which construction firms may find useful in co-creating social value in collaboration with public sector and non-profit organisations. We identify social enterprises as one way of organising operations so that a social or an environmental mission, rather than profitability, is at the heart of the business model. We then discuss creating social value through employment practices. Our focus with regards to employment is on employability, understanding labour markets, and pay as some of the key themes relevant for creating social value as an integral element of socially aware management and operations. We also discuss training and apprenticeships as important ways in which organisations can set up specific, targeted programmes for creating social value, before concluding the chapter with a model: how to achieve social return on investment by embedding social value in strategy and practice.

Partnerships and hybrid collaborations

Traditionally, large national and multinational organisations have tended to be commercially and economically driven, offering very few social benefits (Sinkovics et al., 2015: 345), and they are often thought to take work away from local businesses and communities. This is now changing because it is in tension with the social value agenda that prioritises development of work opportunities for local people and employers, especially in deprived areas, with increasing social and corporate pressure to ensure the benefits remain with, and are returned to, communities (Wilson and Post 2013: 718). Social value is becoming important especially for organisations that undertake local authority work; as highlighted in Chapter 1 many governments are developing social procurement policies (as are many socially conscious private construction clients). Increasingly, legislation requires local governments to implement policies that ensure work is procured in ways that bring benefit to the local communities. Some authorities will use social value contribution to filter and select tenderer submissions: it has been suggested that it is not uncommon for government bodies to place a 10–20 percent weighting on social value (O'Connor 2018), and so each contractor must be able to submit evidence and commitment towards social value. Thus, the pursuit of social good is now becoming part of mainstream business practice (ibid.), on par with concerns about conservation of natural resources and economic profitability; the three pillars of sustainability.

Large organisations can play a useful role in helping communities prosper when they undertake construction work in the area and engage with the local businesses and workforce (Abbott and Allen 2004). Where the temporary construction project organisation goes beyond the link to sourcing the short-term labour force, and in contrast seeks to establish connections with communities, there are reciprocal benefits. Extensive sub-contracting is everyday practice in construction project work, and this offers an opportunity for the larger international and national organisations to engage with local companies. SMEs are often better placed to respond to specific social value goals, often due to their smaller, more local sphere of operation (O'Connor 2018). Due to their community-embedded roots and smaller size, they are often more versatile, have good local knowledge and can engage with many voluntary and community service organisations. Thus, partnerships and hybrid collaborations present a form of organising that helps create social value through market harmony between large organisations and SMEs generating business opportunities for, for example, social enterprises and social businesses, within the supply chain.

Strategic thinking about social value, and organising operations collaboratively at the intersection of public, social and private spheres of economic activity are increasingly commonplace arrangements for delivering value to broad sets of stakeholders (Mandell, Keast and Chamberlain 2016; Quélin, Kivleniece and Lazzarini 2017). Forming new inter-organisational partnerships that bring together governmental, business and non-profit domains as hybrid collaborations aim beyond traditional contract-based partnerships. Research indicates that relational coordination, where focus is on social and work relationships, is vital to social value creation in hybrid collaborations (Caldwell, Roehrich and George 2017). Diverse conceptions of value are integral to understanding the essence, rationale and potential in hybrid cross-sector collaborations; and effective relationships, mutual knowledge and goal alignment enhance the potential for social value creation (ibid.).

Social enterprises and social businesses – with Anna Mimms, MBE

As Loosemore and Higgon's (2015) review of social enterprise in the construction industry notes, social enterprises operate at the intersection of business and society, engaging in commercial activities with the aim of benefiting specific groups of people in the community such as the unemployed, people with disabilities and those form disadvantaged backgrounds. Social enterprises are 'hybrid' organisations which trade in the open market for a social or environmental purpose. They blend economic and social goals, are 'mission-driven' rather than 'profit- driven' and their performance is judged by the difference they make to the communities in which they operate (their social impact) rather than by the profits they generate for private shareholders. Indeed, one of the key characteristics of social enterprises is that the majority of their profits go back to the communities and causes they are set up to serve.

Most commonly, social enterprises are relatively small, local operations that are working to improve a community or support vulnerable people (Social Enterprise UK 2017). They often have a big vision. Thus, social enterprises require a break-the-mould visionary, a leader or an entrepreneur, someone with sufficient passion and creativity to take a risk on doing things differently. Thus, social enterprises make for an ideal 'ideas engine', an opportunity to develop innovative and exciting thinking which can help major contractors more closely connect to the communities in which they work.

Joseph Schumpeter (1947, 1949, 1976) is widely credited for laying the conceptual foundations of entrepreneurship when he described entrepreneurs as the heroic 'wild spirits' who drive the process of 'creative destruction' which lie at the heart of innovation. Research shows that these people have certain attributes in common which include: a sense of purpose; belief, infectious passion, dedication and enthusiasm for their idea; a willingness to take risks and tolerate failure; resilience to bounce-back from failure; humility; intellectual curiosity; optimism; vulnerability; authenticity; generosity; and openness (Muller and Becker 2012; Tjan, Harrington and Hsieh 2012). The growing body of research in social innovation and entrepreneurship is showing that social entrepreneurs tend to also have a particularly strong sense of right and wrong, of altruism and empathy with others who may be less fortunate than themselves, of accountability and responsibility to their communities and of moral outrage against injustice and inequity in society (Dees 2001; Yujuico 2008; Abu-Saifan 2012).[1] For social entrepreneurs wealth creation and profit-making is a means to address a pressing and intractable social or environmental problem, rather than an end in itself.

Sometimes the big vision is simple; for example, to win public sector construction contracts and use them to create local jobs. Whether it is due to economic downturn, such as the 2008 global financial crash, or because of the labour market conditions in certain geographic areas, employing those who are struggling to enter the labour market may require an intermediary, such as a social enterprise, that will secure contracts with large construction projects. The social clauses local authorities commonly include in their tendering processes relate to local labour, apprenticeships and positive local impact. For a large construction company these values may be secondary to the margins delivered to their investors. Social value outcomes are a 'nice to have' with few points of audit or payment relating to their delivery. In contrast, social enterprises work to a social mission. They do business differently delivering innovative solutions to local problems (Social Enterprise UK 2017). The very nature of a social enterprise means that they are more likely to absorb the risks associated with employing entry-level staff; difficult as that may be. They tend to be more forgiving and are less likely to penalise the wages of staff when they inevitably make mistakes, and are more likely to employ people whose complex background and limited tacit knowledge mean that projects are not always as efficient as more commercial outfits.

There are many different types of opportunities to engage with social enterprises in creating social value. Indeed, as we alluded to earlier in the introduction to this chapter, a partnership model within which larger and medium sized construction organisations 'co-create' (Kiser, Leipziger and Shubert 2014: 5) social value in collaboration with a social enterprise means that at project level, construction operations establish a presence in the community that enables lasting relationships to be developed with suppliers and local people. The project partners have an opportunity to lead the way on social value creation, and be a real force for good, for example by procuring through social enterprises.

Social enterprises can be engaged throughout the whole supply chain, anything from trades, recycling firms, stationary suppliers, and media companies, for example. The sectors they operate in are as wide and varied as any other type of business. In addition, social enterprises are more likely to be led by women or ethnic minorities (Social Enterprise UK 2017: 31). In an industry grappling with diversity, partnering with social enterprises is one useful way of empowering different sections of the community to engage with construction operations. By using them in the supply chain, companies can diversify who they procure key business services from, while doing good, playing a key role in supporting the growth of organisations that employ people normally excluded from the construction sector and which are tackling some of society's most pressing problems such as homelessness, unemployment, offending and disability.

Alongside the growth of social enterprises, social businesses are also emerging as an attractive hybrid form of an organisation that utilises market-based approaches to social value creation. A social business is a commercial business with social objectives at its core. It is not a normal commercial business with a corporate social responsibility policy or social conscience. The key is that the business exists for a social purpose, although unlike social enterprises there is no requirement for the majority of profits to be reinvested in that business's cause. People and institutions invest in a social business purely for social purposes, not personal profit. Investors can gradually recoup the money they invest in the enterprise, but do not take any profits beyond that point. Social businesses therefore combine organisational objectives and values associated with for-profit and non-profit activity, which have traditionally been considered contradictory and thus represent powerful social innovation (Wilson and Post 2013). Muhammad Yunus (2007) is a pioneer in social business on a global scale. He was awarded the Nobel Peace Prize for founding the Grameen Bank and inventing the concepts of micro-credit and microfinance. He offers a model for a humane form of capitalism where social businesses do not rely on donations. They are distinguishable from charities in that, although they are social-objective-driven, they operate on a self-sustaining basis and enjoy the potential for growth and expansion (ibid: 23).

Because social enterprises and social businesses are most commonly small organisations, cash flow and access to finance are significant concerns for them (Social Enterprise UK 2017: 41). Issues with cash flow often necessitate those running small businesses to adopt a short-term outlook, which can be detrimental to long-term development, especially in creating and building the connections so essential in securing on-going work opportunities. Social enterprises and social businesses also tend to operate in the most deprived areas (ibid: 14) where clients may experience difficulties with their finance; a problem that is passed on to the SME.

Negotiating work opportunities and continuity of the business are further concerns for social enterprises and social businesses, especially given the project-based nature of work in the built environment. There is often an over-reliance on a single point of contact in a client/contracting organisation whose value system is well-aligned with social value. Changes in personnel can therefore mean discontinuity in work opportunities. Sometimes negotiation of work may feel more like asking for personal favours, rather than a business meeting.

It is then often the single visionary leader who must try and secure the continuity of business whilst also managing the operations. When there is a very close sense of fit between an individual's values and interests and the values and needs of the organisation, it is more likely that the work feels meaningful and purposeful. However, this can also lead to overworking and development of a sense of personal responsibility for the business as well as work, which is often the case for those working in and leading social enterprises (Roper and Cheney 2005: 100). Over time the pressure may result in burnout (Seppala and Moeller 2018).

These challenges do not negate the argument for social enterprise; they have a crucial role to play in creating and delivering social value on a very local level, often in response to a set of specific circumstances, and sometimes for a given period of time only. As the conditions that they were set up to address change, it may be that they are no longer the most efficient vehicle for the required and scalable social value, and an approach embedded within wider society becomes sustainable. Collectively and over time however, social enterprises make a difference in communities across the globe; they help realise and embed change long-term (Hamilton, Crisp and Powel 2016). This creates social value by way of a ripple effect that trickles social value into communities, sometimes indirectly through the wider supply chains. Organisations can also think of creating social value through socially aware employment practices within the organisation, as we discuss in the next section.

Employment and employability

Employment is one of the key pillars of creating social value within communities with a significant potential for change. After years of neglect, this theme has returned to the forefront of the international development agenda, following on the heels of the global financial crisis and its aftermath (Fisher 2014). It is recognised that the social value of jobs includes both the explicit and direct value in exchange, such as pay, and also subjective value, such as sense of purpose and security (Jahan et al. 2016). There is a consensus that unemployment ought to be generally avoided not only because it is detrimental to incomes and demand, but also because it is central to dignity and social cohesion. Employment offers an opportunity for individuals to make a meaningful contribution to society. For organisations seeking to create social value through employment the practical concerns include creating employment opportunities for people from disadvantaged communities, breaking cycles of unemployment and underemployment, equal opportunities, pay, training and apprenticeships.

Diversifying the construction workforce to create social value

The built environment and construction work span a wide variety of different labour markets at both professional and operative levels, although traditionally many groups such as women and people with disabilities have been excluded. This has created a relatively homogeneous workforce and a stereotype image of what the 'ideal' professional or worker in the built environment may look like. Creative analysis of labour markets offers an opportunity for an organisation to create social value through workforce diversification and supporting access of disadvantaged groups of people to worthwhile and rewarding work (Acemoglu 2001).

The jobs people do, whether worthwhile and rewarding or not, are not always determined by their abilities, educational attainment and skills. Class and social mobility have an influence on the kinds of jobs people do (MacLeod 2008). Moreover, women, ethnic minorities, people with disabilities, and ex-offenders tend to occupy a disadvantaged place in many labour markets (Byrne, Clarke and Meer 2005; Kraal, Roosblad and Wrench 2009). This is known as labour market inequality. Labour market inequality refers to the gap between those who have access to worthwhile and rewarding work and those who have low paid jobs with limited prospects, and usually reflects broader concerns about equality within societies. Labour market inequality is heightened when the supply and demand for labour are not in balance. Generally, the demand for labour will be influenced by different stages in the economic cycle (boom/ recession), stages in economic development, patterns of consumption, technological advances, and the strategic decisions managers make about where and how to employ people (for example permanent/casual arrangements). Supply of labour on the other hand is determined by the number of people available for work, their skills and qualifications, and attitude to work/work ethic, mobility (local/global) and expectations.

Over the last two to three decades, there has been a significant reduction in the demand for manual and manufacturing labour while demand for business services and professional workers has more than doubled (Wood 1998; CIPD 2018). This means that although construction labour markets have remained relatively prosperous (CIPD 2018) there are more low-skilled and unskilled workers available in the labour market competing for the manual jobs on offer. Thus, employers can be more selective about who they hire, and it is the disadvantaged groups within labour markets that tend to suffer the most. Social value programmes can be targeted to tackle labour market inequality and increase access to worthwhile and rewarding work, especially for those that have traditionally been excluded from such work in construction (English and Hay 2015; Wright 2015). While location, size of an organisation, market rates of pay, the nature of work, and many other factors will influence the relative balance of worthwhile and rewarding work on offer and who does those jobs, an employer can choose to actively try and tap into specific labour markets. Conducting research on who the disadvantaged groups are in the local labour markets relevant to a project organisation for example is an important first step toward creating social value. Thereafter, targeted recruitment and selection practices can assist the disadvantaged groups' entry to worthwhile and rewarding work.

Indeed, many social value initiatives target labour market inequality by seeking to create opportunities for engagement for a widened pool of people, including women, ethnic minorities, people with disabilities and ex-offenders. As we show in Chapter 2, it is against the law to refuse someone a job because of their gender, or ethnic orientation, or because they have a spent conviction or caution (unless it is because a DBS check shows that they are unsuitable for a specific role). However, in practice, implicit and explicit discrimination is widespread. For example Hughes (2013) found that three-quarters of employers actively discriminate against candidates who declare past convictions. This serves to exclude potentially valuable talent from employment and creates barriers for accessing skills and competencies needed in industry.

The built environment presents a unique opportunity for delivering social value through employment and workforce diversification by recruiting from these marginalised groups. On the one hand, construction work is often the first port of call for disadvantaged groups, for example in employing ex-offenders. The work environment is tolerant of disadvantage and in places culturally diverse (Caplan and Gilham 2005; Loosemore et al. 2010). It is a great local job creator, operating in remote and disadvantaged communities as well in the heart of busy cities. On the other hand, as noted above, the number of women and people with disabilities employed on construction sites tends to be very low (Briscoe 2005; Sunindijo and Kamardeen 2017; Newton and Ormerod 2005, 2013).

At the heart of reconsolidating such opportunities and creating an environment for the employment of a diverse range of workers are organisational policies on equality of opportunity and/or managing diversity and inclusion in the workplace. While some ideas related to equality of opportunity and managing diversity and inclusion in the workplace are broadly the same, i.e. they both focus on fairness and opportunities in employment and training, we concur with a school of thought that sees an important conceptual divide between the two.

Equality of opportunity centres around the protection of groups of people on the basis of protected characteristics that have been identified in law. In the UK, age, disability, gender reassignment, marriage and civil partnership, pregnancy and maternity, race, religion and belief, sex and sexual orientation are the protected characteristics in law, as we identify in Chapter 2. Such protection by law is about a societal level case to treat everyone equally (Kraal, Roosblad and Wrench 2009); an external force driven by legislation. In the UK, the Equality Act also places duty on public sector employers to actively promote equality and foster good relations.

Managing diversity and inclusion in the workplace is a broader concept that relates to the specific individual circumstances of people rather than the characteristics of a group (such as women/men) with an emphasis on harnessing difference for business benefit. Managing diversity and inclusion is about a company's philosophy, whilst equality of opportunity is often more about changing systems and practices. At the level of an individual managing diversity and inclusion means that characteristics beyond those written in law matter; notable examples being class and educational background.

Strategy and practice on creating social value, equality of opportunity, and managing diversity and inclusion in the workplace thus usefully coexist to guide thinking and development of an organisational value base and philosophy together with operational systems and processes. Some social value initiatives may seek to level the playing field, for example by offering additional or specific training programmes to a particular group of people, or working to quotas in recruiting new candidates. When an employer takes steps to help or encourage certain groups of people with different needs, or who are disadvantaged in some way, access work or training, it is called 'positive action' or 'affirmative action'. Positive action is the term commonly used within European countries (Caruso 2003). Affirmative action tends to be used in other countries (ibid).

Positive action and affirmative action are contested terms. There are clear benefits to the minority groups whose position in the labour market is enhanced by such social value initiatives (for example, greater opportunities for employment, decent working conditions, and access to training). Yet, often simply because of confusion and inconsistency relating to the terminology, and lack of information about the nature and purpose of positive/ affirmative action, such initiatives are viewed in a negative light. The European Commission (2009) provides a review of positive action in Europe, Canada, the US and South Africa noting that,

> Positive action measures are widely perceived as politically contentious and require sensitive handling and careful introduction or renewal within organisations. The context for positive action is complex — its development will increasingly require a firm evidence base of best practice to demonstrate that it remains an effective avenue towards progress within pluralist cross-cultures, and that its benefits outweigh the dilemmas it can sometimes raise.

Employers can take positive action to help groups of people when:

- they are disadvantaged in some way in relation to work;
- their participation in employment or training is particularly low; or
- they have particular needs which are different from other people.

Chapters 8 and 9 in Part 2 of our book showcase some good practice in this area in the private and third sectors.

Pay and benefits

Similar to most industries, there is considerable variation in the levels of pay within different parts and levels of an organisation. Executives and professional employees or workers enjoy enhanced pay and rewards packages. Manual and unskilled labour tend to receive limited benefits and minimum pay. In terms of social value, beyond compliance with the legal requirements, current key issues for consideration include differences in the national minimum wage and the living wage, and pay progression barriers for the lowest paid (see Chapter 2 for a discussion of the legal implications of the National Living Wage).

National minimum wage is not statutory in all countries. Nor is the living wage. For example, in countries with strong collective bargaining mechanisms there tends to be industry/sector specific agreements rather than nationally set minimum pay standards.

Where national minimum wage exists, the level is usually calculated on the basis of a comprehensive analysis of the current economic conditions in a given country and therefore what the country could reasonably afford and still compete effectively within the global economy. Consideration of individual circumstances and cost of living do not normally feature in decision-making. Rather, ensuring high levels of employment and minimising unemployment at a national level are a priority. So a national minimum wage is set so that a maximum number of people can remain/ be employed. It is also important to prevent exploitation (CIPD 2014).

Living wage on the other hand is determined on the basis of what an employee may need in order to enjoy a reasonable standard of living within a specific geographic area. The living wage is often significantly better than the national minimum wage. Thus, when considering the social value of employment opportunities for the individuals in question, it is important to understand the difference between national minimum wage and the living wage and the implications of paying one or the other on their personal financial and social circumstances.

In areas where poverty, crime and unemployment are high, employees and workers tend to occupy jobs with very low pay. Individuals within such disadvantaged communities may therefore find it hard to maintain positive perceptions about work and it can be challenging to generate aspirational prospects regarding employment.

The OECD (2018) provides statistics on the levels of national minimum wage in 32 countries. The range is noteworthy. In 2016 the lowest three annual minimum wages reported in US dollars were:

- $1,895.70 per annum in Mexico
- $3,199.20 per annum in the Russian Federation
- $4,753.60 per annum in Brazil

The highest three in the same year were:

- $22,836.10 per annum in Luxemburg
- $22,209.80 per annum in the Netherlands
- $21,967.20 per annum in Australia (closely followed by Belgium, France and Germany)

Analysis of the data on national minimum wages in relation to its power to provide income at a level where work is a feasible route out of poverty highlights alarming variation between different countries (Garnero 2015). At the lowest point, an employee would be required to work almost 80 hours per week to move above a relative poverty line. At the other end of the scale, where the level of national minimum wage level is set high, working 20 hours per week achieves the same.

Hence, statutory systems can only offer a guide, at best, in terms of helping employers to determine appropriate socially responsible pay levels. Simply following national standards may not produce the social outcomes and equity of opportunity an organisation wishes to generate in its global workforce, although on the surface it may look like that.

Another issue of concern is that migrant labour, those workers who relocate to another country in search of opportunities for work (Anderson and Blinder 2017), tend to occupy low skilled job roles (CIPD 2013) and achieve lower levels of pay than their national counterparts (Goodrum 2004; Santoso 2009). This is a cause for concern at the level of the individual worker, and also organisationally. Lower paid migrant workers may not earn enough to support themselves and their families above the poverty line. They frequently suffer from debt bondage as many borrow to migrate; they are required to work very long working hours to compensate for the low hourly pay; they tend to operate in dangerous, dirty and difficult working conditions; and often live in the open or in very poor accommodation with inadequate water and sanitation. In addition, they may experience discrimination and language barriers; they frequently do experience a lack of safety (Santoso 2009); and their families often suffer from an inability to gain access to schooling for their children and an inability to access assistance and government services.

Organisationally unjust pay systems may also lead to difficulties in teamwork and integration of migrant workers on site and feelings of being treated unfairly, which often leads to reduced performance and productivity (Santoso 2009) and even claims for compensation via the industrial tribunal system where experience, level of training, nature of work, and geographical location can't be reasonably argued to contribute towards differential pay. Many employers are keen to employ migrant workers as they come with better job-specific practical skills, work ethic and qualifications; they are better prepared for work, and have more work experience (CIPD 2013). Hence, it makes good business sense to try and retain migrant workers, and fair and reasonable pay offers the means to communicate such intent to the worker.

Where it is not commercially viable to consider pay above the national minimum wage or the living wage, especially for manual and low-skilled work, sometimes explicit communication about the non-financial benefits available helps individuals realise and appreciate their full rewards package. For example, training opportunities and job security may not be registered as part of the pay and benefits package in the mind of a worker who has been out of employment, education and training for long periods of time. Social value can therefore be created through both provision of fair pay and clear communication of the full range of benefits that enhance individuals' position and prospects now and in the future.

Training

Training is often perceived as one specific way of creating social value, an initiative consisting of clearly defined courses, although it can become an integral aspect of an organisation's approach to creating social value. Social groups that have benefited from training can support their community by providing a potential workforce for contractors and the wider supply chain involved in construction activity (Waterhouse, Virgona and Brown 2006). The socio-economic benefits extend to people who do not normally engage in gainful employment and those that have experienced challenges entering into employment; indeed, the number of groups that have proved successful with the long term unemployed, ex-offenders and other disadvantaged groups is growing (for example see the work of Construction Best Practice 2018; Groundwork 2018; and our case studies in Chapter 9).

Training is often a necessary and transformational element of delivering social value within disadvantaged and deprived communities. In such areas school leavers tend to have poor experiences in the education system and thus develop negative perceptions of learning and development and are likely to be found in low-pay, low- skills sectors of the labour market, most at risk of unemployment (Marchington, Wilkinson and Marchington 2008a: 314). Longer serving workers also tend to remain in the low-skill domain and suffer from limited opportunities, and precarious work. Where support, training and work opportunities come together, a positive sense of community can develop and long-term synergies and prospects can be created (ILO 2011). Such synergies mean that organisations benefit from a better-educated and motivated workforce that is able to respond to opportunities offered by construction activity. Individuals benefit from prospects for gainful employment and opportunities to use and develop their skills and knowledge. This in turn holds the potential to uplift self-esteem and engender a sense of purpose and a positive contribution to society. At a national level, a skilled and motivated workforce provides opportunities for economic growth and social inclusion (Nilsson 2010; ILO 2011; Asadullah and Ullah 2018). In developing countries, training can help reduce large scale poverty (Nilsson 2010).

In sum, training is a way of creating social value at three different levels – societal, organisational, and individual– achieving a myriad of benefits at all levels. As the International Labour Organisation (2011: 4–5) points out, training:

- empowers people to develop their full capacities and to seize employment and social opportunities;
- raises productivity, both of workers and of enterprises;
- contributes to boosting future innovation and development;
- encourages both domestic and foreign investment, and thus job growth, lowering unemployment and underemployment;
- leads to higher wages;
- when broadly accessible, expands labour market opportunities and reduces social inequalities.

Training can include longer-term and broader development of skills and knowledge, for example through education. Training can also consist of providing opportunities for learning off-the-job and/or on-the-job through specific activities which focus on skills provision or upskilling. Courses and classroom-based training are generally the easiest to record and cost, and often the most visible form of training within the workplace. They offer protected time for learning and the chance for participants to exchange ideas with each other and learn from shared experience in a structured setting (CIPD 2017a). In collaboration with further and higher education providers organisations can offer varied opportunities and allow learners to acquire formal qualifications. Such a functionalist perspective holds a positive view of education and training as a way of helping maintain society by socialising young people into the values of achievement, competition and equality of opportunity while providing key skills for organisations.

Building on extensive research worldwide, Harrison (2009) advocates a partnership model to delivering training in practice; an approach aligned with co-creating the social value ideology discussed earlier in this chapter (see Kiser, Leipziger and Shubert 2014). Key to such value-added training is a shared determination by the trainers, learners, and sponsors to co-create learning and value (ibid). This helps ensure training is based on shared ownership and not simply delivered to the learners, which in turn engenders pride in learning and developmentand facilitates advancement. Ideally, training adds value to the business, the trainers/learning providers, and the individuals. Employing a systematic approach, such as the six-stage training cycle (Harrison 2009: 184) or the ADDIE (Analysis, Design, Development, Implementation and Evaluation) model (Kurt 2018), facilitates the design and implementation of effective training interventions. Both tools offer a somewhat linear process, however, it is rare that they will ever operate in an entirely sequential fashion. Instead, several of the stages may be conducted in parallel, or some may fall out of sequence at times.

The six-stage value-adding training cycle (Harrison 2009: 183–227) consists of:

1. establishing the partnership;
2. integrating planning and evaluation;
3. identifying training and learning needs;
4. agreeing learning principles and strategy;
5. designing and delivering training;
6. monitoring and evaluating outcomes.

Construction organisations and project teams frequently engage in informal and on-the-job type learning and training activities, yet more systematic and strategic thinking about human resource development is often seen as someone else's responsibility (Loosemore, Dainty and Lingard 2003: 253). There are many reasons for this which include the fragmented nature of the industry and the large number of small-scale organisations which do not have the time and resources to invest in training. There are also structural problems with national training systems and apprenticeship schemes and competency standards to guide nationally accredited systems that provide transferability of skills and pathways of progression for those who wish to progress their careers. There is also the problem of an ever-changing minefield of skills-related policies and initiatives confusing employers and employees (Moehler, Chan and Greenwood 2008). Apprenticeships are one such prominent training intervention closely connected to the social value agenda.

Apprenticeships

Generally apprenticeships refer to a workplace-based learning system that generates opportunities for employment and training for young people, and they are a particularly powerful mechanism in providing a pathway to work for those who have no formal qualifications because of their disadvantaged background. More specifically, contemporary descriptions of apprenticeships tend to refer to work-based learning programmes, which combine paid employment or work experience with on-the-job and off the-job learning.

National systems range from provision of vocational education and training through full-time schooling in vocational colleges that offer teaching and training, to a dual system based primarily on workplace-based apprenticeship learning. The dual system commonly includes part-time vocational schooling and training in firms that accept apprentices.

The German apprenticeship training system is a renowned exemplar of the dual system. It has been considered as one of the pillars of German economic competitiveness. Apprenticeship in Germany includes a combination of structured training in business and technical skills advanced in vocational school. Even relatively small firms consider training and development as important aspects of management and employment relations in Germany (De Massis et al. 2018). A significant number of employers believe in the central philosophy of life-long learning and the implementation of the German apprenticeship system is extensive. Around two thirds (60 percent) of young people in Germany progress to employment through the apprenticeship scheme (Jacobs 2017). The social value in the system lies in its foundations and tasks aimed at upgrading students learning motivation, their problem-solving capacity, self-esteem and their moral character (Spariosu and Spariosu 2017) together with the direct impact on employment and employability at societal level. The system offers a direct path from schooling and training to employment, and eases the transition period (Zhao 2018).

Most of the German apprenticeship trainees join three-year programmes that either mix company-based and in-school training, or consist solely of school-based training with work placements (Jacobs 2017). In the first version, companies pay for most of the training and the apprentice wages; in the second, state governments pay for the training in schools (ibid). Both culminate in nationally recognised qualifications.

Despite much interest across the globe, given contextual differences between countries, it is often difficult or impractical to try and implement the German model elsewhere. Shared apprenticeship schemes, a standards based apprenticeship framework, and employer-led apprenticeship standards offer valuable alternatives to the German model.

In the UK, the CIPD (2017b, 2017c) provides guidance on apprenticeships for employers. They advocate apprenticeships as a unique way to 'grow your own' by combining on-the-job training in an organisation with off-the-job learning, and providing employers with an effective way of growing their skills base (CIPD 2017c). The apprentice's learning takes place in context and provides a real understanding of the working world, combining practical skills with theoretical knowledge.

The core elements of many apprenticeship-training schemes focus on developing:

Transferable skills

These skills are variously referred to as functional/key skills, core skills, or essential skills, but all these terms describe a core set of skills that people need in today's workplaces. They include English, maths, and information and communications technology (ICT) using practical applications.

Competence

The competence (technical skills) aspects of the apprenticeship are usually based on National Occupational Standards and are completed in the workplace.

Knowledge

The knowledge part of the apprenticeship covers the technical knowledge and theory that is relevant to the practical skills an apprentice will develop in their job role.

Personal learning and thinking skills or wider key skills

These are the skills that help an apprentice to succeed at work as independent enquirers, creative thinkers, teamworkers, reflective learners, self-managers, and effective participators.

(CIPD 2017b)

The construction industry has always utilised apprenticeships and on-the-job training extensively. Naylor, Raiden and Morgan (2009) report on the South West Wales Shared Apprentice Scheme, which is a partnership arrangement between the Welsh Assembly for Wales, local employers and learning providers, such as local further education colleges. This project began in 2007 with the piloting of two shared apprenticeship schemes in Carmarthenshire; one in the engineering sector (24 places) and the second in construction (48 places), and was extended to a regional model in 2014 (CITB 2014).

The unique overall aim was for a group of SMEs to collaborate and share a number of apprenticeship places, overcoming the structural problem associated with industry fragmentation of burdening one small firm with the need to support an apprentice by itself. Their joined-up approach and partnership with the learning providers created the opportunity for the local authority to adopt a more wide-spread strategic planned approach towards construction apprenticeships. This is in sharp contrast to the previous ad-hoc appointment of trade apprentices on a three-year employment cycle. The project has been recognised as good practice, with 'additionalities' highlighted as a particularly strong feature. For example, bricklayers learn additional modules in plastering, thus developing 'a multi-skilled apprentice that will play a key role in delivering WHQS in Carmarthenshire' (Kiely 2008). Similar 'additionalities' have also been delivered in other enhanced shared apprenticeship models including:

- An Australian construction model which incorporates financial literacy and business skills in their final year (Daly 2007).
- A Northern Ireland employer-led gas sector apprenticeship scheme that offers extra modules enabling apprentices and their firms to offer clients a 'one-stop-shop' for all central heating services (Moore 2008).

Another shared apprenticeship scheme that closely resembles that of the South West Wales construction scheme has been successfully running in Sydney, Newcastle, Canberra and Adelaide in Australia. Master Builders (2018) operates Apprenticeship Group Training Schemes and places with different contractors for varying periods over the duration of their training. The apprentices are employed by Master Builders for the duration of their apprenticeship. The main philosophy behind the rotation of the apprentices through various employers is to broaden their skills. The role of the host employer is critical in providing on the-job experience. In Australia, both main contractors and sub-contractors are included. In Wales, the scheme utilises local authorities to provide a range of skills development opportunities, with an emphasis on repairs and maintenance.

The shared apprenticeship schemes offer job security for the apprentices and the receipt of pay while they learn on the programme. There is also an expectation that the apprentices will be directly employed by the local authority or major contractors and SMEs on completion of their training.

In Ireland, a 'standards based' apprenticeship framework, aimed at replacing the emphasis on 'time served with common standards' (O'Connor, 2003) thrived during the Celtic Tiger, an economic boom from the mid-1990s to the late-2000s (Murchadha and Murphy 2016). This model was focused on supporting apprentices' achievement of a particular standard, for example a good knowledge of maths, through provision of off-the-job training and workplace-based learning activities. As such, less emphasis is given to serving a specific period of time on a programme, as is commonly the case with schemes in Germany, UK and Australia where three years service has been the norm.

New, employer-led, apprenticeship standards are now replacing the common standards in Ireland and the UK. These schemes put employers in the driving seat as they help design the apprenticeships and manage the contracts with training providers. The apprenticeships range in duration from two to four years, and lead to nationally recognised awards and qualifications.

The overall objective of the new apprenticeships standards is to ensure that apprenticeships are truly employer-led. The standards are therefore designed by employers (trailblazer groups) to meet their needs, the needs of the respective sector and the economy more widely (Apprenticemakers 2018). The standards give an occupational profile and specify the knowledge, skills and behaviours required to successfully undertake a specific occupation, and to operate confidently within a sector (see the Institute for Apprenticeships 2018 for published standards).

There are also some innovative local apprenticeship schemes delivering extensive social value. For example, the 'St Basils Birmingham Live Work' programme operating in Birmingham in the UK is part of a network of self-help community-led housing ventures. It was established to provide apprenticeships for young homeless people, or those at risk of homelessness. Uniquely, the scheme provides on-site shared accommodation for the apprentices in partnership with Keepmoat Regeneration, who complete refurbishment works on derelict and void properties. The scheme provides an opportunity for young people to live and work without recourse to their welfare benefits.

Managing aspirations

Collectively, the above mentioned ways of creating social value through employment, including workforce diversification, pay, training and apprenticeships, hold the potential to influence change in people's perceptions and aspirations regarding work. The meaning of work can develop from being seen as a straightforward performative/ productive activity ('putting in the hours for pay') to a rewarding and enriching experience (Cartwright and Holmes 2006) with the potential to role model and uplift local and societal value systems.

Creating positive perceptions and aspirations within groups of workers who are traditionally excluded from the built environment regarding work, employment and careers offers a window of change within disadvantaged communities and helps break cycles of long-term underachievement and apathy. MacLeod (2008) offers powerful insights into the process of social reproduction: how class structure is reproduced by a combination of forces, such as social status, parenting, schools, and work; and how working-class young people do not even aspire to middle-class jobs. It follows that class inequality is reproduced within society. Prospects for socioeconomic advancement are doomed before they even get started.

To achieve change in people's perceptions and aspirations regarding work, organisation/project specific social value schemes and initiatives may look at avenues for engaging people with the wider community around them. Offering work placements to young adults and school children from disadvantaged backgrounds is a way local authorities often encourage and develop a sense of aspirational employment within local communities.

Other initiatives may be focused on:

* providing career advice and information for young people on specific careers;
* offering curriculum support to schools, with employers (for example contractors) sharing knowledge and expertise about their work, and the nature of jobs;
* targeting hard to reach groups;
* creating specific opportunities and offering support for the long-term unemployed, and those not in education, employment or training (NEETS).

The 'Inspiring and Creating Social Value' toolkits from Croydon (2012) and Blackpool Council (2013) local authorities in the UK showcase their approach to engaging with the local communities in creating social value and managing aspirations.

The CIPD (2014) reports on an 'hourglass' phenomenon that has been seen in the UK, the US and most other developed economies, whereby there are now fewer transparent opportunities for progression within organisations. Their forecasts suggest this trend will only accelerate over the coming decade or two. The squeeze on traditional 'middle tier' jobs has meant that, all too often, fewer opportunities are available for people from disadvantaged backgrounds to progress from the bottom tier in the workplace to the middle, let alone the top. Thus, even where people have the aspiration to 'get on' or 'reach higher' they will find that the link between hard work and reaching the next rung of the ladder is elusive.

Importantly, creating social value therefore becomes more than an entry-level programme of encouraging people from disadvantaged backgrounds into work. It is crucially also about progression and development, showcasing longer-term prospects and creating higher level aspirations.

Embedding social value in strategy and practice

Over the longer term, creating social value may progress beyond specific and targeted initiatives and become embedded in the ways in which an organisation operates. As social value becomes embedded in corporate activity and strategy, it becomes easier to legitimise the multiparty benefits. As the benefits to the community and environment are realised, the renewed relationship with the community should return added value to the organisations that engage. Relationships brings multiple benefits, including more effective recruitment of local resource, better deployment of services through improved engagement, ability to manage any inconvenience suffered locally at the point of disruption and during works, reduced public health issues, and improved safety and security (Buck, Harrison and Brookes 2015). The knock on effects of social value activities in communities can regenerate community outlook, bringing services, infrastructure, employment, health, wellbeing, training, collective responsibility, increased resilience and social inclusion. When social activities and initiatives are tailored, the impact can be considerable and the knock-on effects broad, with the health and wellbeing of communities benefiting (Alen and Alen 2015). The key principle of social value is that for investments in everyday organisational operations, such as supply chain management, and employment and training, there are social opportunities that can bring added value.

Social value initiatives and programmes that offer very specific impact (for example targeted training events or short-term employment trials) are of benefit, but may offer limited long-term impact if the overall prospects remain poor. Equally, from an organisational perspective, securing long-term benefits of specific stand-alone initiatives can be challenging given that, for example, employer investments in apprenticeship training are sometimes a risky strategy because well-trained and competent workers are very attractive to competitors and prone to poaching, and the training firm may lose its investment (Mohrenweiser, Zwick and Backes-Gellner 2018). Even where there are large capital investments, and organisations and projects are able to deliver sports facilities, community halls, and health centres to areas in need, benefits beyond the one-off social value investment into the community can be difficult to sustain. The facility operational costs should not be overlooked as these may be higher than the initial capital expenditure (Dobson et al. 2013). If the local communities are not able to support the operation and maintenance costs, then the amenities can fall into disrepair and may close. Focusing on whole-life costs is key as demonstrated by the case study in Chapter 7.

Securing holistic long-term commitment is a major challenge, as sustainable social value takes time to embed the necessary change in the community. Thus, social value initiatives may be seen as being fragmented, reactive and/ or 'paying lip service'. Contractors, local authorities and other stakeholders may be saying 'but we do all of that' in response to requests for investment into social value programmes. Those working to a compliance model tend to do the same thing; they fulfil the requirements in planning regulations and other aspects of law. This leads to a pedestrian approach to creating social value: few apprentices, basic training and employment conditions, a cursory lick of paint in the community centre, limited volunteering service; and hence, missing out on the real opportunity to create social value.

Summary

This chapter has focused on creation of social value, opening with discussion about partnering, hybrid organisational forms and social enterprises as ways of co-creating social value at a strategic, organisation level. Collaboration with social enterprises is highlighted as way of creating social value through the construction supply chain. We have also discussed ways of creating social value within construction organisations, primarily through employment practices. Our focus was on understanding labour markets, managing aspirations, and pay, together with training and apprenticeships as important ways in which organisations can set up specific, targeted programmes for creating social value.

Key points

- Partnerships and hybrid collaborations present a form of organising that helps create social value through market harmony between public, private, and third sector organisations and throughout the supply chain.
- Social enterprises are often small, local operations with a big vision, working to improve a community or support vulnerable people.
- Many social value initiatives target labour market inequality by seeking to create opportunities for engagement for a widened pool of people, including women, ethnic minorities, people with disabilities and ex-offenders.
- When an employer takes steps to help or encourage certain groups of people with different needs, or who are disadvantaged in some way, access work or training, it is called 'positive action' or 'affirmative action'.
- Payment of a living wage is an effective mechanism for helping people work their way out of hardship and poverty.
- Training is a way of creating social value at three different levels –societal, organisational, and individual– achieving a myriad of benefits at all levels.
- Apprenticeships are a particularly powerful mechanism to provide a pathway to work for those who have no formal qualifications because of their disadvantaged background.
- Social value holds the potential to influence change in aspirations regarding work: perceptions of work can develop from being seen as a straightforward performative/ productive relationship to a rewarding and enriching experience with the potential to role model and uplift local and societal value systems.
- Embedding social value into strategy and practice helps return added value and achieve multiparty benefits over long-term.

Note

1 See also Messick and McClintock 1968; Kelley and Stahelski 1970; Kelley and Thibault 1978; Griesinger and Livingston 1973, on the psychology of social value orientation.

References

Abbott, C. and Allen, S. (2004) Exploring how regional construction activity can be supported by community benefit initiatives. In: Khosrowshahi, F. (Ed.), *Proceedings 20th Annual ARCOM Conference, 1–3 September 2004, Edinburgh, UK*. Vol. 1, 353–361.

Abu-Saifan, S. (2012) Social entrepreneurship: definition and boundaries. *Technology Innovation Management Review*, February, 22–27.

Acemoglu, D. (2001) Good jobs versus bad jobs. *Journal of Labor Economics*, 19(1): 1–21.

Alen, M. and Alen, J. (2015) Using the Social Value Act to reduce health inequalities in England through action on the social determinants of health. Available online at: www.gov.uk/government/uploads/system/uploads/attachementdata/file.460713/1aSocial ValueAct-Full.pdf.

Anderson, B. and Blinder, S. (2017) *Who Counts as a Migrant? Definitions and their Consequences*, 5th edn. University of Oxford: The Migration Observatory. Available online at: www.migrationobservatory.ox.ac.uk.

Apprenticemakers (2018) The small business Q&A: what are apprenticeship trailblazers? Available online at: http://apprenticemakers.org.uk/the-small-business-qa-what-are-apprenticeship-trailblazers.

Asadullah, M.A. and Ullah, A.Z. (2018) Social-economic contribution of vocational education and training: an evidence from OECD countries. *Industrial and Commercial Training*, 50(4): 172–184.

Austin, E.G. (2011) Good job, bad job – how much should we care about the quality of newly created jobs? *The Economist*, 14th September.

Blackpool Council (2013) Inspiring & Creating Social Value in Blackpool: Social Value Toolkit. Blackpool Council. Available online at: www.blackpool.gov.uk/Business/Working-with-the-council/Documents/Inspiring-and-creating-social-value-in-Blackpool.pdf.

Briscoe, G. (2005) Women and minority groups in UK construction: recent trends. *Construction Management and Economics*, 23(10): 1001–1005.

Buck, D., Harrison, D., and Brookes, C. (2015) Local action on health inequalities: Using the Social Value Act to reduce health inequalities in England through action on the social determinants of health. Public Health England, Practice resource: September 2015. Available online at: www.gov.uk/government/uploads/system/uploads/attachment_data/file/460713/1a_Social_Value_Act-Full.pdf.

Byrne, J., Clarke, L., and Meer, M.V.D. (2005) Gender and ethnic minority exclusion from skilled occupations in construction: a Western European comparison. *Construction Management and Economics*, 23(10): 1025–1034.

Caldwell, N.D., Roehrich, J.K., and George, G. (2017) Social value creation and relational coordination in public-private collaborations. *Journal of Management Studies*, 54(6): 906–928.

Caplan, A.S. and Gilham, J. (2005) Included against the odds: failure and success among minority ethnic built-environment professionals in Britain. *Construction Management and Economics*, 23(10): 1007–1015.

Caruso, D. (2003) Limits of the classic method: positive action in the European Union after the new equality directives. *Harvard International Law Journal*, 44(2): 331–386.

Cartwright, S. and Holmes, N. (2006) The meaning of work: the challenge of regaining employee engagement and reducing cynicism. *Human Resource Management Review*, 16: 199–208.

CIPD (2013) *The State of Migration – Employing Migrant Workers*. London: Chartered Institute of Personnel and Development.

CIPD (2014) *Pay Progression: Understanding the Barriers for the Lowest Paid*. London: Chartered Institute of Personnel and Development.

CIPD (2017a) *CIPD Factsheet: Learning Methods*. London: Chartered Institute of Personnel and Development.

CIPD (2017b) *Apprenticeships: An introduction Factsheet*. London: The Chartered Institute of Personnel and Development.

CIPD (2017c) *Apprenticeships That Work: A Guide for Employers*. London: The Chartered Institute of Personnel and Development.

CIPD (2018) *Labour Market Outlook*. London: Chartered Institute of Personnel and Development.

CITB (2014) Shared Apprentice Schemes in Wales. Available online at: www.citb.co.uk/news-events/wales/shared-apprenticeship-schemes-wales/.

Construction Best Practice (2018) *Creating Opportunity for an Ex-offender*. Graham Construction. Available online at: https://ccsbestpractice.org.uk/entries/creating-opportunities-for-an-ex-offender.

Croydon (2012) *Inspiring and Creating Social Value in Croydon: A Social Value Toolkit for Commissioners*. Croydon Council. Available online at: www.croydon.gov.uk/sites/default/files/articles/downloads/socialvalue.pdf.

Daly, G. (2007) Industry: financial literacy pilot shifts into high gear, educators and trainers. *Financial Literacy Network Newsletter*, No. 3, October 2007.

Dees, J.G. (2001) *The Meaning of Social Entrepreneurship*. Durham, NC: Duke University, Fuqua School of Business, Center for the Advancement of Social Entrepreneurship.

De Massis, A., Audretsch, D., Uhlaner, L. and Kammerlander, N. (2018) Innovation with limited resources: management lessons from the German Mittelstand. *Journal of Product Innovation Management*, 35(1): 125–146.

Dobson, D.W., Sourani, A., Sertyesilisik, B. and Tunstall, A. (2013) Sustainable construction: analysis of its costs and benefits. *American Journal of Civil Engineering and Architecture*, 1(2): 32–38.

English, J. and Hay, P. (2015) Black South African women in construction: cues for success. *Journal of Engineering, Design and Technology*, 13(1): 144–164.

European Commission (2009) *International Perspectives on Positive Action Measures – A Comparative Analysis in the European Union, Canada, the United States and South Africa*. Luxembourg: Office for Official Publications of the European Communities.

Fisher, A. (2014) Social values, employment and human development: beyond economic utilitarianism. *United Nations Development Programme Human Development Reports*. Available online at: http://hdr.undp.org/en/content/social-values-employment-and-human-development-beyond-economic-utilitarianism.

Garnero, A. (2015) *Minimum Wages across OECD and EU Countries, OECD Mutual Learning Programme-Learning Exchange, London, 11 September*. Available online at: file:///C:/Users/hmd3raideab/Downloads/Minimum%20wages%20across%20OECD%20and%20EU%20countries%20-%20Andrea%20Garnero%20(2).pdf.

Goodrum, P.M. (2004) Hispanic and non-Hispanic wage differentials: implications for United States Construction Industry. *Journal of Construction Engineering and Management*, 130(4): 552–559.

Griesinger, D.W. and Livingston, J.W. (1973) Toward a model of interpersonal motivation in experimental games. *Behavioral Science*, 18(3): 173–188.

Groundwork (2018) MSSTT Changing Places Changing Lives. Available online at: www.groundwork.org.uk/Sites/msstt/pages/blue-sky-msstt.

Hamilton, A., Crisp, R. and Powel, R. (2016) *Overcoming Deprivation and Disconnection in UK Cities*. Joseph Rowntree Foundation. Available online at: www4.shu.ac.uk/research/cresr/sites/shu.ac.uk/files/overcoming-deprivation-disconnection-uk-cities.pdf.

Harrison, R. (2009) *Learning and Development*, 5th edn. London: CIPD.

Hughes, E. (2013) Why employers should stop discriminating against ex-offenders. Personnel Today, 21 October. Available online at: www.personneltoday.com/hr/why-employers-should-stop-discriminating-against-ex-offenders.

ILO (2011) *A Skilled Workforce for Strong, Sustainable and Balanced Growth: A G20 Training Strategy*. Geneva: International Labour Organization.

Institute for Apprenticeships (2018) *Bricklayer*. London: Institute for Apprenticeships. Available online at: www.instituteforapprenticeships.org/apprenticeship-standards/bricklayer.

Jacobs, R. (2017) Germany's apprenticeship scheme success may be hard to replicate. *Financial Times*, 21 April.

Jahan, S., Jespersen, E., Mukherjee, S., Kovacevic, M., Abdreyeva, B., Bonini, A., Calderon, C., Cazabat, C., Hsu, Y.-C., Lengfelder, C., Luongo, P., Mukhopadhyay, T., Nayyar, S., and Tapia, H. (2016) *Human Development Report 2016*. New York: United Nations Development Programme.

Kelley, H.H. and Stahelski, A.J. (1970) Social interaction basis of cooperators' and competitors' beliefs about others. *Journal of Personality and Social Psychology*, 16: 66–91.

Kelley, H.H. and Thibault, J.W. (1978) *Interpersonal Relations: A Theory of Interdependence*. New York: Wiley.

Kiely, J. (2008) *Maximising Benefits Across Wales*. Welsh Housing Quality Standard Plus, Savills Housing Consultancy.

Kiser, C., Leipziger, D., and Shubert, J.J. (2014) *Creating Social Value: A Guide for Leaders and Change Makers*. London: Routledge.

Kraal, K., Roosblad, J., and Wrench, J. (eds) (2009) *Equal Opportunities and Ethnic Inequality in European Labour Markets*. Amsterdam: IMISCOE Reports, Amsterdam University Press.

Kurt, S. (2018) ADDIE Model: Instructional Design, Educational Technology. Available online at: https://educationaltechnology.net/the-addie-model-instructional-design/.

Loosemore, M. and Higgon, D. (2015) *Social Enterprise in the Construction Industry: Building Better Communities*. London: Routledge.

Loosemore, M., Dainty, A., and Lingard, H. (2003) *Human Resource Management in Construction Projects: Strategic and Operational Approaches*. London: Spon Press.

Loosemore, M., Phua, F., Dunn, K., and Ozguc, U. (2010) Operatives experiences of cultural diversity on Australian construction sites. *Construction Management and Economics*, 28(2): 88.

MacLeod, J. (2008) *Ain't No Makin' It*. New York: Routledge.

Mandell, M., Keast, R., and Chamberlain, D. (2016) Collaborative networks and the need for a new management language. *Public Management Review*, 19(3): 326–341.

Marchington, M., Wilkinson, A., and Marchington, L. (2008) Vocational education, training and skills. In: Marchington, M. and Wilkinson, A. (eds), *Human Resource Management at Work*, 4th edn. London: Chartered Institute of Personnel and Development, pp. 309–341.

Master Builders (2018) *Master Builders Apprentices*. SkillsOne. Available online at: www.skillsone.com.au/vidgallery/master-builders-apprentices.

Messick, D.M. and McClintock, C.G. (1968) Motivational bases of choice in experimental games. *Journal of Experimental Psychology*, 4: 1–25.

Moehler, R.C., Chan, P.W., and Greenwood, D. (2008) The interorganisational influences on construction skills development in the UK. In: Dainty, A (ed.), *Procs 24th Annual ARCOM Conference, 1–3 September 2008, Cardiff, UK*, pp. 23–32.

Mohrenweiser, J., Zwick, T., and Backes-Gellner, U. (2018) Poaching and firm-sponsored training. *British Journal of Industrial Relations*. doi.org/10.1111/bjir.12305

Moore, R. (2008) *First Ever Employer-Led Engineering Apprenticeship Launched in Northern Ireland*. Skills Focus, Energy and Utility Skills, Issue 8.

Muller, T. and Becker, L. (2012) *Get Lucky. How to Put Planned Serendipity to Work for You and Your Business*. San Francisco: Jossey-Bass.

Murchadha, E.Ó. and Murphy, R. (2016) Rethinking apprenticeship training for the construction industry in Ireland. In: Chan, P.W. and Neilson, C.J. (eds), *Proceedings 32nd*

Annual ARCOM Conference, 5–7 September 2016. Manchester, UK: Association of Researchers in Construction Management, pp. 373–382.

Naylor, G., Raiden, A., and Morgan, A. (2009) Constructing apprenticeships: transforming through sharing. In: Dainty, A.R.J (ed.), *Proceedings 25th Annual ARCOM Conference, 7–9 September 2009, Nottingham, UK, Vol. 1*, pp. 635–644.

Newton, R. and Ormerod, M. (2005) Do disabled people have a place in the UK construction industry? *Construction Management and Economics*, 23(10): 1071–1081.

Nilsson, A. (2010) Vocational education and training – an engine for economic growth and a vehicle for social inclusion? *International Journal of Training and Development*, 14(4): 251–272.

O'Connor, K. (2018) Social value metric means contractors have to change. Construction Manager, 11 February.

OECD (2018) Real Minimum Wages. Available online at: https://stats.oecd.org/Index.aspx?DataSetCode=RMW.

Quélin, B.V., Kivleniece, I., and Lazzarini, S. (2017) Public-private collaboration, hybridity and social value: towards new theoretical perspectives. *Journal of Management Studies*, 54(6): 763–792.

Roper, J. and Cheney, G. (2005) The meanings of social entrepreneurship today. *Corporate Governance: The Inernational Journal of Business in Society*, 5(3): 95–104.

Santoso, D.S. (2009) The construction site as a multicultural workplace: a perspective of minority migrant workers in Brunei. *Construction Management and Economics*, 27(6): 529–537.

Schumpeter, J.A. (1947) The creative response in economic history. *Journal of Economic History.* 7: 149–159.

Schumpeter, J.A. (1949) Economic theory and entrepreneurial history. In: Wohl, R.R. (ed.), *Change and the Entrepreneur: Postulates and the Patterns for Entrepreneurial History.* Cambridge, MA: Harvard University Press.

Schumpeter, J.A. (1976) *Capitalism, Socialism and Democracy.* London: Routledge.

Seppala, E. and Moeller, J. (2018) 1 in 5 highly engaged employees is at risk of burnout. Harvard Business Review, 2 February.

Sinkovics, N., Sinkovics, R.R., Ferdous Hoque, S., and Czaban, L. (2015) A reconceptualization of social value creation as social constraint alleviation. *Critical Perspectives on International Business*, 11(3–4): 340–363.

Social Enterprise UK (2017) *The Future of Business – State of Social Enterprise Survey 2017.* London: Social Enterprise UK.

Spariosu, T. and Spariosu, B.B. (2017) Macroeconomic effects of vocational education and training. In: Stojkovic, D., Bozovic, M., and Randjelovic, S. (eds), *Economic Policy of Smart, Inclusive and Sustainable Growth.* Belgrade, Serbia: University of Belgrade, Faculty of Economics, pp. 153–176.

Sunindijo, R.Y. and Kamardeen, I. (2017) Work stress is a threat to gender diversity in the construction industry. *Journal of Construction Engineering and Management*, 143(10).

Tjan, A.K., Harrington, R.J. and Hsieh, T. (2012) *Heart, Smarts, Guts and Luck: What it Takes to be an Entrepeneur and Build a Great Business.* Cambridge, MA: Harvard Business Press.

Waterhouse, P., Virgona, C., and Brown, R. (2006) *Creating Synergies: Local Government Facilitating Learning and Development through Partnerships. A National Vocational Education and Training Research and Evaluation Program Report.* Available online at: https://files.eric.ed.gov/fulltext/ED493942.pdf.

Wilson, F. and Post, J.E. (2013) Business models for people, planet (& profits): exploring the phenomena of social business, a market-based approach to social value creation. *Small Business Economics*, 40: 715–737.

Wood, A. (1998) Globalisation and the rise in labour market inequalities. *The Economic Journal*, 108: 1463–1482.

Wright, T. (2015) New development: can 'social value' requirements on public authorities be used in procurement to increase women's participation in the UK construction industry? *Public Money & Management*, 3(2): 135–140.

Yujuico, E. (2008) Connecting the dots in social entrepreneurship through the capabilities approach. *Socio-Economic Review*, 6(3): 493–513.

Yunus, M. (2007) *Creating capitalism world without poverty –social business and the future of capitalism*. New York: Public Affairs.

Zhao, Z. (2018) Modern apprenticeship as an effective transition to working life: improvement of the vocational education system in China. In: Pavlova. M., Chi-Kin Lee, J., and Maclean, R. (eds), *Transitions to Post-School Life*. Singapore: Springer, pp. 51–66.

5   Social value assessment

*Ani Raiden, Martin Loosemore, Andrew King
and Chris Gorse*

*In Chapter 1, we defined what we mean by the term social value. We also discussed
examples of what it looks like in practice. We noted that the concept of social
value is context dependent, has different potential in different projects, that it may
change over time, affect some groups more than others and be seen by different
groups in significantly different ways. In this chapter we discuss how all of these
issues need to be accounted for in any assessment of social value/impact. We outline the ongoing debate surrounding the contentious issue of how social impact is
measured and describe a structured methodology and set of principles for doing
so in the context of the built environment.*

The case for social value assessment

It is not possible to quantify and monetise all forms of social value and it is for
this reason, that we have decided to use the term 'assessment' rather than 'measurement' in the title to this chapter. While we acknowledge that the accepted terminology in the social value literature is to use the term 'measurement', we feel
the term assessment is more appropriate because the term measurement tends to
be associated with quantification and monetisation. We agree with Nicholls et al.
(2012: 8) who note that, 'Every day our actions and activities create and destroy
value: they change the world around us. Although the value we create goes far
beyond what can be captured in financial terms, this is, for the most part, the only
type of value that is measured and accounted for'.

We define social impact assessment as the process by which those in the
built environment account for all types of value that their businesses, projects
or programmes create, beyond just economic impacts to include social, environmental, cultural and health impacts at both a community and individual level.

Being able to assess social value is crucial for a number of reasons:

- It enables managers to make more informed decisions about the full range of
  impacts they wish to create beyond just financial impacts.
- It enables managers to identify where the greatest social value is being created
  in a business or project and in its supply chain so that they can focus on value-
  adding activities and suppliers to maximise their impact.

- It is crucial in communicating effectively with key stakeholders (for example staff, customers, funders, investors, communities).
- It allows managers to incentivise positive behaviour by linking social value outcomes to rewards and sanctions.

In simple terms, the assessment of social value/impact can be predictive and forward looking (a forecast impact assessment) or retrospective (an evaluative impact assessment). Crucially however, to assess the social value of a social initiative/program accurately one must be involved throughout its life (inception, through design, implementation and close-out), since understanding the goals of a programme are crucial to understanding its impacts, many of which may take time to eventuate.

In reading this chapter, it is important to appreciate that the assessment of social value is a young and emerging field of study where there remain a range of unresolved controversies and challenges. These include:

- Inconsistent terminology.
- Questions around validity and reliability (do measures of social value measure what they purport to measure and can they be replicated).
- Inconsistent assessment methodologies, approaches to data collection, measurement and reporting (no universally accepted international standards currently exist).
- Assessments varying depending on time invested and who undertakes the process.
- Philosophical, ethical and practical criticisms of utilitarian techniques like SROI (which seek to monetise outcomes, reduce assessments to a single ratio and assume that the correct action is the one that maximises economic utility).
- The need to accommodate and consider multiple stakeholder views.
- A lack of legitimacy and trustfulness in what is reported (biased and selective reporting... mostly good news... failures are relatively under-reported).
- Skills and knowledge in assessing social outcomes vary considerably (internationally recognised qualifications do not yet exist and there are insufficient skills and knowledge about what to measure, how to measure, what metrics to use and how to analyse and interpret the data).
- Professionalisation and exploitation of the field (resulting in overly complex and expensive assessments beyond the reach of small businesses like social enterprises).
- The lack of independent certification/auditing and reporting standards.
- The large gap between theory and practice in most social value assessments (due to skill, time and cost constraints).
- Poor communication of results.
- Assessments not being linked to strategy or being used to direct strategy (senior managers not seeing it as a priority or beneficial – adopting a tick-box, compliance-based approach).

- Culturally insensitive approaches to measurement which do not allow assessments to reflect how stakeholders perceive social value from their cultural context (notions of social value differ from culture to culture).

While social impact measurement has become more important as the public sector moves towards evidence-based policy and outcomes-based procurement, it remains one of the most contested issues in policy, research and practice. Research in this area is still in its infancy and while many good reports, reviews, best practice guidelines and toolkits have been produced, there remains a lack of conceptual clarity about social impact measurement with very few robust, comprehensive and empirical studies to draw from. As Pritchard et al's (2013) research shows, these serious limitations mean that it is often difficult for managers and policymakers to argue with any confidence that the impact of their policies or activities have made a real difference in the communities in which they operate.

The discussions above raise many important questions in developing our skills in measuring the social value of construction projects. First, it is clear that the complexity of social value and the lack of standardisation and consistency used in assessments makes comparisons difficult. This is especially the case when they are monetised and all combined into one overall ratio as in the social return on investment (SROI) approach. Many argue that such ratios grossly over simplify the many nuances, variables and complexities in fully understanding the social value created by any intervention. Second, there are problems when comparing social value measurements over different periods of time and different projects and programmes, since value means different things at different times to different people as discussed in Chapter 1. Third, there are legitimate questions of whether construction professionals are currently equipped to measure the social impact of the industry's activities. As Close and Loosemore (2015) note, the majority of practitioners in the construction sector have traditionally excluded communities from the construction process, seeing them as a risk rather than an asset, and have not developed the skills or inclination to effectively understand their needs and aspirations. If we are to understand the social impact of the built environment, then it is critical that we address these educational and attitudinal deficiencies so that construction professionals can appreciate that the community's perceptions of value are not only shaped by the extent of potential benefit or harm that we as experts might calculate, but by the way in which people in that community interpret it.

To achieve this, effective and meaningful community engagement, rather than tokenistic community engagement, must be recognised as a crucial part of the social value assessment process. As Loosemore and Phua (2011) argued, construction managers should appreciate that scientific and technical assessments of risks and opportunities associated with construction projects by experts are relevant only to the extent that they are integrated into individual community perceptions. In other words, it is society that dictates social value, not experts, and the creation of social value must be built on an intimate understanding of what value means to those communities.

Measuring social value in practice

While there is as yet no singular commonly accepted standard for measuring social value, there is no shortage of guidelines available for those who want to try. Some of the most well respected and known include: Nicholls et al. (2012), CSI (2014) G8 (2014) ICAEW (2015) NCVO (2013) NPC (2014) and NSW (2015).

The GECES (2014) review of the many approaches to social impact measurement provides an excellent critique of the field and although it is aimed at social enterprises, and is primarily driven by the needs of funders, investors and policy makers, there are some useful findings and recommendations for contemplation in the context of the built environment.

First, in seeking to arrive at a common approach for social impact assessment across the EU, GECES (2014) concluded that there is no point in developing an approach or setting standards that are excessively rigid, costly to meet or impractical in the information required from the businesses involved and its beneficiaries. To be practical and realistic, social impact assessment must be:

- Set in the context of, and be designed to support, the decisions to be made, and the learning expected to be gained from it.
- Proportionate and appropriate to the purpose, data and resources available to undertake an assessment.

Importantly, GECES (2014) notes that financial proxies and financial indicators for measurement (such as those frequently used in SROI) should only be used if they add value to the key stakeholders' viewpoint.

Second, GECES (2014) recommended that any approach must provide easily decipherable and understandable information to enable real decision-making around the direction of resources to improve social impact. This applies to all stakeholders, not just funders, and needs to include policy makers and customers who use the product or service. Third, GECES (2014) argues that many quantitative indicators that are commonly used to measure social impact fail to capture, and can even misrepresent and under-value, some essential qualitative aspects of impact, and instead need to be balanced by a multi-method approach that adopts a range of tools and data (both quantitative and qualitative) that are able to capture the story of change that has occurred in the lives of people affected by a project or programme.

We agree with these general principles because if clients, funders or other stakeholders on the construction sector imposed a rigid approach to social impact assessment, then it is highly likely that the indicators chosen would be problematic and reduce innovation in the design of social programmes to meet community needs. More specifically, it is likely they would be misaligned, both with the constantly changing and varied needs of the communities in which buildings and infrastructure are built, and with the many organisations in the supply chain that deliver them. In our view, any approach needs to reflect a particular sector's diversity, which in construction means meeting the needs of the many large, as well as small, contractors and consultants, operating across a wide range of building types and geographic project areas and social needs. Furthermore, the recommendation that any assessment approach must be proportionate to the time and resources available, size of the organisation measuring and the risk and scope for the intervention being delivered, is particularly relevant to the built environment given the large numbers of small firms that dominate the industry and the lack of understanding of social value that currently exists. Devising a rigid set of social impact indicators in a top-down and one-size-fits-all fashion would not only ignore the diversity of the built environment, but would make it difficult to fairly and objectively capture the huge variety of potential impacts created at multiple levels and sit uncomfortably with the key need for proportionality. Indeed, the wrong indicators could provide a perverse incentive, driving behaviours in the wrong direction away from the effective delivery of valuable social outcomes, or even lead organisations to 'game the system', organising themselves to maximise their achievements against inappropriate indicators, rather than to achieve the greatest social impact in response to community needs.

While GECES (2014) argues that a standardised approach to assessing social impact would be inappropriate, it does recognise the need for agreement on common principles of good assessment. To this end, it argues that any social assessment exercise should be:

1. Relevant: related to, and arise from the outcomes it is measuring.
2. Helpful: in meeting the needs of stakeholders', both internal and external.
3. Simple: both in how the measurement is made, and in how it is presented.
4. Natural: arising from the normal flow of activity to outcome.
5. Certain: both in how it is derived, and in how it is presented.
6. Understood and accepted: by all relevant stakeholders.
7. Transparent and well-explained: so that the method by which the measurement is made, and how that relates to the services and outcomes concerned, are clear.
8. Founded on evidence: so that it can be tested, validated, and form the grounds for continuous improvement.

These principles are similar to those that are widely cited as being the seven core principles of social impact measurement by peak bodies such as Social Value International (2016), which have themselves been based on those underlying the more established fields of social accounting and auditing; sustainability reporting, cost benefit analysis and financial accounting and evaluation practice. These are listed below, and by applying these principles consistently it is possible to create a valid, reliable and credible account of social value for a programme, which can stand up to scrutiny. The seven core principles of social impact assessment are:

1. Involve stakeholders – Stakeholders are those people that experience change because of the project or programme. Key stakeholders should be involved in any assessment of social value since they are best placed to describe the change.

2. Understand what changes – The changes that result from the project or programme for each key stakeholder should be evaluated and described and supported by evidence as attributable to the project or programme being assessed. The changes evaluated should be immediate, intermediate and long-term and both positive and negative and should include those that affect both 'primary' stakeholders (in the target population) and 'secondary' stakeholders (in the people delivering the activities and in the target population's families and wider communities).

3. Value the things that matter – Value what matters to key stakeholders. The assessor should not impose their own view of value on stakeholders but seek to assess social impact from their perspective. This principle is important for all stakeholders, but especially when assessing traditionally disempowered groups or groups which are not the same culture, background or generation as the assessor. While a wide range of changes might occur, some will be more important than others and not all can be included in social impact assessments for practical reasons. So, it is important to focus on those that matter most.

4. Only include what is material – only assess changes that can be 'evidenced' by primary and secondary data collected using a mix of qualitative and quantitative methods and triangulated and cross-referenced from a range of perspectives to ensure that an unbiased, fair, balanced and accurate account of change is produced. To do this rigorously is time-consuming and costly and requires a deep understanding of interpretivist research methods which treat respondents as 'meaning makers', recognise the socially constructed and culturally relative nature of social value and which are able to respect those perspectives.

5. Do not over-claim – Only claim the value of changes that can be attributed to the activity being assessed taking into account various *counterfactuals* which include: deadweight (what would have happened anyway); drop-off (reducing benefit over time); attribution (what else could have contributed to the change); displacement (what other benefits does the intervention displace/push aside); and substitution (replacement of other gains). This will require independent assessment by qualified assessors and comparison to baselines and past trends in the indicators being used to measure change collected before the project or programme was implemented.

6. Be transparent – Don't hide anything and be prepared to be accountable for any changes claimed. Report both positive and negative changes, reveal all data sources and limitations and communicate to all stakeholders.

7. Verify the result – Verify the results with key stakeholders and ensure appropriate independent verification, although organisations providing verification are currently few in number and since there are no internationally agreed standards, these verifications are themselves open to question.

The social value assessment process

The approach to social value assessment described in the remainder of this chapter is based on a synthesis of approaches to social impact assessment and measurement by leading organisations around the world (as cited above) and on our own experience of undertaking social impact measurement in practice in the built environment. The process is an integrated five step approach, which seeks to address a key problem with current methods; they fail to reflect the crucial connection between the upfront planning, design and implementation of a social programme/intervention and the assessment of its social impact in practice. We believe that too often, social value assessment is portrayed as something to do after a programme is finished, effectively an afterthought. The five steps in our approach are as follows:

1. planning and programme development;
2. developing a theory of change;
3. developing a measurement framework;
4. implementing the programme and assessing and monitoring social impact;
5. reporting to stakeholders, learning and improving.

Step 1 – Planning and programme development

In being true to a bottom-up approach, the community and its social needs and priorities should represent the foundation of any social programme and correspondingly any subsequent assessment of its impact in that community. So, the first step in any social value assessment should be to undertake a community needs assessment to understand community social needs, priorities, strengths and weaknesses. By taking time to learn about the community in which one is building, one can discover the most relevant opportunities for a project to direct finite resources most effectively in areas where it is needed to close the gap between where the community is now and where it needs to be in the future. A community needs assessment will also help to identify the desired outcomes against which success will be measured and the boundaries of the assessment in terms of stakeholders who will be included in the process (for example, individuals, groups, communities and families) and how deeply into the community any social value assessment will penetrate.

Conducting a community needs assessment

The Department of Planning and Community Development (2011) in Australia's Victorian Government has developed a useful 'Communities Mapping and Analysis Methodology' to drive its social procurement initiatives. This presents a useful six stage framework for analysing and identifying training and employment opportunities for targeted disadvantaged communities which can be generated through the social procurement process and the 'readiness' of the local community to respond to future social procurement opportunities. The steps in this framework are as follows:

1. Identify key unemployment issues or disadvantaged groups in the local community – identify the key unemployment issues, and disadvantaged

groups, that exist within the local community that might be assisted by social procurement strategies. Examples include: unemployment amongst specific community groups (for example youth groups, indigenous people, newly arrived migrants and people with disabilities). It is wise to focus on one group and issue which is achievable within time and resource constraints and past experience and which is strategically aligned to organisational values and goals.

2. Map and forecast local investment projects/jobs and skills demand – this step aims to identify, map and forecast investment and related skills demand in the context of current and future construction projects. The aim is to ensure that the project or programme in question contributes to any gaps in employment opportunities that may arise in the future and not replicate opportunities already in the pipeline. At an individual project level, it involves looking at the skills forecast over the life of a project according to its construction programme, but should also include other projects within and outside an organisation, providing opportunities to collaborate with other organisations that create 'sustainable' job opportunities to address social issues beyond the life of individual projects.

3. Map local job readiness and skills development providers/services/programmes – This step aims to identify and map existing providers, services and programmes which are currently addressing the identified community needs. The information generated by this exercise will avoid potentially overlapping or counterproductive activities that could undermine existing programmes and potential social impacts and establish a provider database to guide potential collaborative relationships in meeting community needs. Cross-sector collaboration is essential since no one organisation can solve social problems alone.

4. Map existing local skills/skills gaps relevant to forecast demand – Taking account of the identified community needs, the skills needed over the life of a construction project or number of projects, and the corresponding existing organisation and programs catering for them, this step aims to identify gaps that need to be addressed by the project's social procurement activities.

5. Analyse the opportunity to plan for social procurement activities – This step asks whether there are social procurement solution/s consistent with the organisation's CSR goals, values and internal procurement policies and priorities, which have the capacity to meet the identified gap in community needs. It also asks who are the key stakeholders and how will they be engaged in taking the social procurement process forward.

6. Annually review future investment/social procurement opportunities – This stage involves identifying any new future investment opportunities that may create additional demand for skills and jobs and any other developments with existing providers that might influence the opportunity for social procurement.

The six-step process described above sounds straight forward, but as recent research into communities impacted by construction project shows, they are highly dynamic entities, comprising multiple and often deeply conflicting groups, with widely differing needs and aspirations that also change over time and over the life of a project, through planning, design, construction and operational phases (Teo 2009). Close and Loosemore (2015) show that these characteristics make community consultation challenging in the construction sector and that these nuances of community structures are lost on most construction professionals who make the process of community consultation costly, burdensome and time-consuming and is an area where they have little experience and skills. Importantly, in the built environment, community consultation is seen to be the domain of urban planners who have developed detailed principles and techniques to interact with communities during the early, pre-construction phases of a construction project. Once the project starts on site, any community concerns are assumed to have been resolved early on, although in reality it is only when construction starts on site that many in the community realise what impact it will have on their lives and then become vocal. Whilst some researchers such as Tam and Tong (2011) have developed useful typologies and frameworks for managing stakeholder expectations, and fostering trust between stakeholders and construction projects, few researchers have explicitly singled-out communities as a stakeholder to be better understood, typically bundling all stakeholders into a singular cohesive group.

Despite these current limitations in community consultation practices within construction, an understanding of community needs and of existing agencies and organisations (not-for-profit, government or private) delivering those needs is the foundation of any effective social programme and social value assessment. Even if one is already involved in a community, a community needs assessment can reveal changes in priorities and additional opportunities for creating social value and it can also help establish important new relationships which will help bring about change. Indeed, even the task of undertaking a social value assessment can help build valuable relationships and encourage community members to actively participate in a social programme.

An effective community needs assessment should be undertaken genuinely and with an open mind and to give voice to marginalised and disempowered groups who might not have been given the opportunity to express their needs in the past. The community consultation literature reminds us that in dealing with marginalised groups, cultural sensitivity is essential to ensure that programmes are developed in ways that are consistent with a people's and community's cultural framework and perceptions of social value, which may differ considerably from an assessor who is external to that group and community (Airhihenbuwa 1995). However, some communities are not always willing to participate in community consultation efforts. For this reason, a large part of community engagement practice highlights the need for trust, empowerment and capacity building by providing communities with the skills, knowledge, tools and authority to enable them to make effective decisions on critical social issues (Fawcett et al. 1995).

This may all sound obvious, but it is surprising how many times the social value assessment process starts with what managers think the community wants, which dooms the programme to failure (at least in the community's eyes) from the very start. Having a few conversations with selected people in the community who are easily accessible is not an effective way of finding out what the community needs, and data should also be collected from a range of different stakeholder groups reflecting the diversity of the community you are building in and using a wide range of qualitative and quantitative data from both primary and secondary sources. To save resources and time, it is often a good idea to consult the various specialist support agencies/organisations that represent community groups for information about the main challenges for their cohorts. However, the credibility, capability and compatibility of organisations with which one might need to collaborate will also need to be established since some may be far more effective than others in meeting community needs. They may also have very different views on how to meet those needs and there are often community politics to manage, particularly in indigenous groups where different tribes may exist. Partnering with the wrong organisation can seriously undermine even the most well-designed social programmes.

Practical stakeholder consultation

Stakeholder consultation lies at the heart of community needs assessment and from a practical perspective, Burby (2001) found that effective stakeholder consultation involves a number of key decisions relating to: objectives; timing; participants; techniques; and information provision.

OBJECTIVES

Many community consultations are ineffective because the *objectives* driving the process are not clearly formulated. In a social programme, possible objectives might include:

- Compliance with social procurement requirements.
- Compliance with social investor/funding requirements.
- To genuinely discover community needs, priorities, expectations, desires, existing strengths and weaknesses and gaps which need addressing.
- To discover existing initiatives and organisations working in the community to ensure the programme complements rather than overlaps with existing efforts.
- To develop collaborative links with existing organisations working in the community to address community needs.
- Educating and informing stakeholders about what is being done in a social programme.
- Building a collaborative culture to mobilise a supportive constituency of stakeholders.
- Understanding stakeholder perceptions and preferences about the structure and outputs of a programme.
- Empowering stakeholders to influence social outcomes and structure of a social programme.

The nature of the objectives driving community consultation and ultimately the shape of a social programme and the data collection and metrics that are used to measure social impact. For example, in the case of meeting investor, funder or compliance requirements, there may be specific groups in the community which will need to be involved, there may be specific social outcome targets to achieve and there may even be specific methods of measurement, metrics and reporting requirements which are required. Finally, in line with the basic social assessment principles of proportionality, it is important that any assessment of community needs should be proportionate with the resources, time and scale of the project or programme being considered. This will determine the boundaries of the assessment process in terms of the range of needs measured, the number of stakeholders consulted and the period and frequency over which the assessment will take place.

TIMING

Decisions about *timing* are linked to the objectives and resources underlying the consultation process. For example, if resources are limited and the objective is simply compliance, then the frequency of meetings will be strictly dictated by funding agreements, laws and regulations. Similarly, if the objective is simply to inform stakeholders of a project rather than meaningfully gain their views, the consultation process might be limited to public hearings at the end of the project. However, if the objective is to genuinely understand community needs in the development and design of a social programme, then the interaction will need to be far more frequent. Burby (2001) found that large dividends in engagement with programmes resulted from early stakeholder participation in decisions.

PARTICIPANTS

Given the breadth of issues raised by a construction project, the complexity of potential communities affected and thus the number of *participants* involved, the task of consultation can be overwhelming. If every stakeholder were consulted individually then many projects would simply be unviable and never go ahead. Burby (2001) found that decision makers who indiscriminately embark on widespread consultations are less likely to be effective than those who adopt a stakeholder management strategy that identifies important stakeholders and focuses resources on them. In the context of social impact assessment, a stakeholder is any person who experiences 'material change' as a result of a project or programme (Nichollset al. 2012). A 'material' change is one which is significant to stakeholders and directly attributable to the project or programme being developed and assessed. Ideally, any stakeholder who is likely to experience (or has experienced) 'material change' should be included in the community consultation process.

However, in reality, given the many potential impacts of construction projects in the communities in which they are built, and the resource and time constraints they are built under, it is not possible to involve and consider every stakeholder in the community. For this reason, it may be useful to refer to the numerous stakeholder engagement tools that have been developed in the business management literature to help managers decide who should be involved in decision-making processes. For example, Mitchell et al. (1997) proposed a framework of stakeholder engagement based on:

- the stakeholder's power (ability to prevent you achieving your goals);
- the legitimacy of the stakeholder's claims (community expectations around their right to be consulted); and
- the urgency of the stakeholder's claims on the organisation (how rapidly they can influence your interests).

In contrast, van Oosterhout and Kaptein's (2008) approach to classifying stakeholders is based on whether a project manager can maintain mutually beneficial relationships with a stakeholder; when this is not possible then they should be excluded from the consultation process. More recently, Littau (2015) adopted Winch's (2004) categorisation of the stakeholders along the three dimensions of power, interest and attitude, organising them into six categories with associated project stakeholder management strategies:

1. Acquaintances (keep them informed with a transmit only communication style).
2. Sleeping Giants (sleeping till other actors, normally negative ones — awake them for having their claims considered. Managers should act proactively to engage them for supporting the project).
3. Irritants (Interested in social and environmental aspects, clear and transparent communication is essential).
4. Friends (project managers should use them as confidants).
5. Saboteurs (power derives from other stakeholders like media or governments. Project manager may change their attitude by providing voice to their claims and using clear and transparent communication. If this is not deemed possible, managers should gain other stakeholder's support to reduce their power).
6. Saviours (these are key players the project manager should pay attention to and keep on side).

However, we caution against the use of such frameworks. As Loosemore and Phua (2011) argue, such frameworks tend to treat community consultation as a process of manipulation and risk mitigation rather than genuine engagement and they can also encourage managers to invest in those groups which hold more power, legitimacy and urgency (typically investors and shareholders) at the expense of those who are traditionally disempowered in society, such as indigenous communities. This is a potential problem that is also highlighted by Social Value International (2016: 5) who recognise that social value 'is often invisible because it relates to outcomes experienced by people who have little or no power in decision making'.

TECHNIQUES

Numerous models and *techniques* of community consultation have emerged out of the vast and long-standing body of research in this area – much of it in the urban planning domain. Although nearly 50 years old, Arnstein's (1969) ladder of citizen participation is arguably the best known framework since it shows how different consultation approaches affect and are affected by balances of power between society and business.

On the bottom rungs of Arnstein's eight-step ladder are 'manipulation' and 'therapy' (non-participation), which are aimed more at educating or curing stakeholder concerns rather than involving them in a decision. Barnes (2002) argues that this approach is underpinned by a belief that the community is irrational, uneducated and misinformed and cognitively unable to make reliable decisions about the impacts of business on their lives. The next three runs on Arnstein's ladder are 'informing', 'consultation' and 'placation' (tokenism) where the community is given a 'voice' in decision-making, but not the authority to ensure their views will be acted on. Barnes (2002) argues that the underlying assumption here is that experts know best and that the subjective perceptions of risk and opportunity held by communities are inferior to expert knowledge and scientific facts and figures. In contrast, in the upper-most steps of Arnstein's ladder, 'partnerships', 'delegated power' and 'citizen control' provide communities with a real opportunity to influence decision-making processes. Barnes (2002) argues that here community perceptions of risk are seen as critically important, regardless of whether they align or not with expert assessments. At this level, there is a recognition that perceptions of risk and opportunity associated with construction projects are socially constructed and that community perceptions of risk and opportunity are as legitimate as expert facts and figures. As Brewer (2013) shows, at this level community participation involves a process of collective social learning between multiple stakeholders who are given appropriate knowledge, power and institutional support to contribute meaningfully within a conducive policy and regulatory environment.

Like all categorisations, Arstein's model has also been criticised. In particular, Tritter and McCallum's (2006) critique asserts that the model is too inflexible and that in reality, community participation is not as hierarchical and linear as Arnstein's ladder suggests. Also, while the ideal of citizen empowerment is implied, there is little discussion of how that power is best exercised or recognition that different types of problems and stakeholders and situations need different levels of empowerment and types of participation and that this can change during the participatory process itself. While Arstein's model has been criticised for being inflexible (Tritter and McCallum 2006) and while other ladder-type models have been produced to address these concerns, such as Burns, Hambleton and Hoggett's (1994) 'Ladder of Citizen Power' and Wilcox's (1999) 'Ladder of Participation', Tritter and Macullum (2006) acknowledge that Arnstein's ladder of citizen participation remains the touchstone for those involved in community consultation, particularly those involving large infrastructure projects.

INFORMATION PROVISION

Whatever approach one takes to community consultation, Burby's (2001) work shows that that the more *information* exchanged between the community and construction project professionals, the more likely it is that a project or programme would be supported in practice and therefore deliver the social outcomes which the community needs. To this end, there are a range of standard methods that are used to communicate with those people who are deemed to be legitimate community stakeholders. These can be found in a multitude of published and online resources around community needs assessments and consultation going back over many years, which in turn draw insights from the vast body of research on social science research methods (see for example, Witkin and Altschuld 1995; Altschuld 2010; Davidson 2005; Watkins, West Meiers and Visser 2012; Rotary 2015; National Resource Center 2010; Our Community 2017).

Broadly speaking these tools include:

- *Community/town hall/public meetings*: A forum, which brings together members of a community to ask questions, discuss issues, voice concerns, and express preferences for community priorities. They should be led by a skilled facilitator and/or leading community member, often backed by a panel of experts and the success of these events depend on whether key stakeholder groups can attend, how well the facilitator organises the event, defines its objectives and controls the discussion, giving all groups a chance to speak. If there are cultural or language barriers then interpreters may be needed and the event should be held in a neutral location to avoid potential bias in the proceedings.

  The advantage of community is in accessing many people from potentially diverse background using limited resources and time. There is a vast amount of social data generated in many forms (verbal, visual, relational, behavioural and emotional), which can provide rich qualitative information about community needs. However, without clear ground rules and effective facilitation, such meetings can be difficult to control and become emotional and unfocused. Furthermore, the quality of interaction may be constrained by the location and timing of the event, which may affect accessibility to certain groups. There may also be underlying social norms, inequalities and power dynamics in the community which may prevent the concerns of marginalised groups being voiced. Finally, attendance is difficult to control, meaning that one cannot guarantee that those who attend are representative of the demographic make-up of the community.

- *Surveys*: Surveys can be administered electronically or face-to-face and have the advantage that they are a low-cost way of accessing large numbers of people over wide geographic areas. The advantage of on-line surveys is the anonymity and potentially large sample sizes they can provide. Surveys can produce large amounts of quantitative data that can be repeated and analysed using descriptive and inferential statistics to reveal trends and correlations.

Their anonymous nature can also be useful for gaining information about sensitive issues that people may not feel comfortable speaking about in an open forum such as a town hall meeting. However, the data produced by surveys is typically quite sterile of social content and surveys are not effective at collecting qualitative information, which are often central to understanding community social needs and impacts. Furthermore, response rates to surveys are often disappointingly low, it is often difficult to know who has responded and self-selection can bias results. Whilst surveys look deceptively simple, effective surveys that elicit unbiased responses need considerable skill to develop and sampling strategies need to be carefully considered to ensure a high response rate from a representative portion of the wider population.

- *Interviews*: Interviews can range from structured to unstructured and can be undertaken face-to-face or remotely using communication technologies such as phones and skype. The advantage of unstructured interviews is the spontaneity that they allow the interviewer in chasing-up and responding to unexpected themes and leads around social needs and impacts that emerge during the interaction. The personal relationship developed during interviews can also help to build trust when discussing personal and sensitive issues that are often central to understanding social needs and impacts. Interviews can also be quite effective when dealing with disadvantaged populations, which may not have the literacy skills to complete surveys and the resources to attend town hall meetings for example. The downside of interviews, especially in a social impact assessment context, is that they are time consuming, sample sizes are small and they require considerable skill in asking questions and establishing the rapport needed to elicit information about social needs and impacts.
- *Focus groups*: Focus groups involve small groups of people engaging in a carefully structured dialogue around a specific issue – normally guided by a set of pre-determined questions asked by a skilled facilitator. Like town hall meetings they produce a rich array of data in many forms, not only in what is said, but in the behaviours and interactions between respondents which can be observed. In addition to verbal discussion, techniques such as community mapping can be used to facilitate and record discussion where focus group participants draw a map of their community, marking certain points of importance guided by specific questions around social challenges, needs and priorities. Whilst powerful, focus groups do not provide reliable information at an individual level and considerable skills are needed to facilitate an effective focus group discussion. Careful preparation is essential to ensure that the focus group reflects the structure of the community being studied and that the group's membership is structured to avoid 'group-think' and destructive conflict whilst at the same time producing a diversity of views. The best ideas are likely to arise from interactions between people with many different positions. However, some community groups might not be comfortable in close proximity interactions, focus groups can be time-consuming for respondents and

there are certain skills and attributes needed to participate effectively that need to be considered by facilitators when selecting participants.

- Technology is increasingly being used to engage communities in projects and contemporary techniques involve interactive websites, social media, e-consultation, multimedia displays, deliberative polling and tele-voting (Troast 2011).

  Whatever information provision and consultation technique is used, some stakeholders may not be equipped to assimilate and understand the information being provided. In other situations, there may be confidential issues that one party may not wish to divulge. The issue of community consultation and needs assessment is therefore not just one of access and quantity, but of content, context and trust. To overcome these problems, stakeholder relationships may need to be nurtured in numerous ways, stakeholders may have to be trained to interpret the information being provided and the information may need to be provided in a number of different forms and ways for it to be understood properly. Building trust with the community may be especially important if previous construction projects have caused harm.

Given all of the above difficulties and complexities, it is not surprising that most construction companies contract-out the community consultation process to a specialist consultant, in a similar way to letting a trade package (Raiden, Dainty and Neale 2006). Cleland (2007) argues that this allows someone with specialist skills to focus on community relationships. However, as Winch et al. (2007) argue, this can be a costly exercise and often portrays to the public a lack of care by the company who is outsourcing their responsibility to the community.

Step 2 – Developing a 'theory of change'

Having identified the community stakeholders you are aiming to serve and their social needs and priorities, the next step is to design your programme to meet these needs. This is done by developing a 'theory of change', which is an evidence-based conceptual framework depicting how the programme will achieve its objectives by visually linking programme inputs, activities, outputs, outcomes and impacts. A theory of change should also articulate any assumptions and enablers (internal and external), which need to happen to allow the theory of change to work as planned. As New Philanthropy Capital (NPC 2014; 2014b) says, the advantage of developing such a theory is that it forces a focus on key inputs and activities that lead to intended outputs, outcomes and impacts. This enables the most effective use of limited resources by allowing them to be focused on activities that add the greatest social value.

Theories of change (sometimes called impact maps or logic models) are normally developed collaboratively in a workshop, based on a combination of secondary data derived from empirical research (which shows that certain interventions lead to certain outcomes), experience of how an intervention has worked in the past (backed up by previous impact studies and research) and primary data from potential beneficiaries (for example surveys, interviews and focus groups, as discussed). The process of collaboratively developing a theory of change is valuable in itself. Not only does it help to forge trusting and collaborative relationships, which are likely to be key to the successful delivery of a project or programme, it also builds a common understanding and sense of shared ownership of the intervention by those who are responsible for bringing it about. Also, by forcing people to think deeply about the causal links relating certain activities to certain outcomes, hidden assumptions by different parties, which may have undermined an intervention, can be revealed and resolved.

Developing a theory of change (despite the term) is not an academic exercise but is critical to effective social impact assessment. In essence, social impact assessment can be seen as a process of collecting data to test the theory of change, which in essence is a conceptual representation of the programme as it is intended to work. The learning that happens in the process enables you to refine a theory of change into a more robust framework to inform future interventions. Over time, as it is tested over and over in practice in different contexts, the theory of change should become more refined, valid and robust. This refined theory of change can then be used to inform the design of similar programmes in the future with greater confidence that they will work as intended. Crucially, a detailed theory of change is the foundation of any effective social impact assessment process since it identifies the significant input, activities, outputs, outcomes and impacts that need to be assessed.

As with most things in impact assessment, there is no standard way to depict a theory of change. Theories of change can be simple or complex and this tends to vary depending on the preferences of those who develop them and on how complex the social programme is. Although many theories of change are depicted in diagrams, some are presented in tabular format. However, it should be noted that a diagram or table is only a representation and a summary of the process of thinking through and describing in full how a project or programme should work. The most important requirement of any good theory of change is that it is easy to understand, that it is supported by evidence and that it depicts clearly how a social programme will work along with any underlying assumptions.

NPC (2014; 2014b) describes four common formats for presenting a theory of change: The CES (Charities Evaluation Services) Planning Triangle, logic model, outcomes chain and written narrative. In describing and illustrating these different approaches, we have provided below an example of a real-life work-integrated training programme being put in place by a construction company as part of its social procurement strategy. This has been designed to get ex-offending youth in the vicinity of a new construction project back to work by giving them supervised paid work experience, training and mentoring.

The programme runs every six months on a continuous basis for the duration of the project provides support to ex-offenders before, during and after transition into employment. Not only does it provide work experience but it provides an opportunity to learn, practice and develop the personal attributes, competencies and skills required to work in the built environment. On completion of the course participants received a Nationally Accredited construction safety qualification which allows them to work on construction projects in various trade and professional roles.

The programme involves partnerships with a range of other organisations such as the Probation Service who refer ex-offending youth to the programme, construction subcontractors mentoring the youth and supervising their work experience, a charity providing local support and coordination and a local higher education college providing course delivery and qualifications.

Candidates are first screened through interviews to assess their readiness to undergo the programme. Those deemed not suitable are referred to other sources of assistance which better match their specific abilities and needs. For selected candidates, the programme is run over six months with a structured safety training programme of four weeks at the start which focuses on construction safety but also work-readiness, confidence building, self-esteem, relevant life skills, literacy and numeracy. The programme then provides certified candidates with five months of supervised and paid work experience in at least three trade areas of their choice. One-to-one mentoring is provided for each candidate by the principal contractor to provide workplace support, advice and assistance when needed. The wages are paid by the principal contractor. Post programme support is provided to help the candidates who are successful seek full-time employment.

The following example is presented as the basis for illustrating the four common formats of theories of change as described by NPC (2014a; 2014b).

The CES Planning Triangle

The CES Planning Triangle (see Figure 5.1) is a simple visual tool developed by Charities Evaluation Services that defines a programme's goals (intended long-term outcomes), intermediate outcomes (what needs to change for beneficiaries to achieve the final goals) and the activities that need to be undertaken to achieve the intermediate outcomes. It is best to start with the desired impacts and work backwards ensuring that each activity has a direct link to one or more of the intermediate outcomes, and each outcome needs to have links to the social impacts. If links cannot be found between activities, outcomes and impacts, then questions should be asked about whether they need to be included.

Logic models

A logic model (see Figure 5.2) is a more detailed version of the CES Planning Triangle and is constructed in the same way by working backwards from the final stated goals. They show the relationships between social inputs, activities, output, outcomes and impacts and can be made more detailed by linking inputs, activities, outputs, outcomes and impacts using numbering and colour coding. They can also be used to reflect inputs and outcomes for different stakeholders. However, whilst

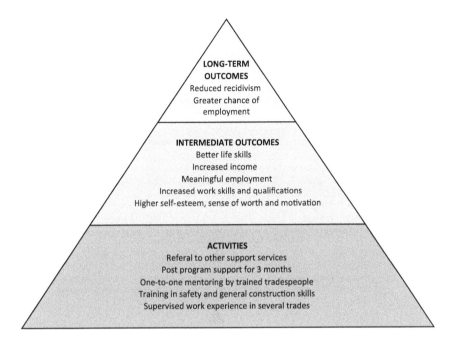

Figure 5.1 The CES Planning Triangle

they provide more detail than the CES Planning Triangle, the limitation of logic models and the CES Planning Triangle is that they over-simplify the causal complexities of how many social programmes work, giving the impression of a neat transition from input, to activity, to output, to outcome and then finally to impact.

In reality, social impacts do not occur in such a linear way and these limitations can be overcome by using outcomes chains and narrative theories of change.

Outcomes chains

Outcomes chains (see Figure 5.3) depict causality between activities and outcomes (short-term, intermediate and long-term), articulating in flow chart format how and why change occurs and the conditions needed for success. An outcomes chain shows the sequence in which outcomes are planned to occur and the key causal processes and enablers representing the critical path to success. Without enablers, a theory of change cannot happen, so explaining the enablers is important. As in the case of logic models and planning triangles, the process of designing an outcomes chain begins by agreeing the final goal (long-term outcomes one wishes to achieve) and then working backwards, thinking about cause and effect, to identify the intermediate outcomes, immediate outcomes and activities from which they emerge. Every activity and outcome should eventually link to at least one of your aims. However, mapping causality is challenging because many things

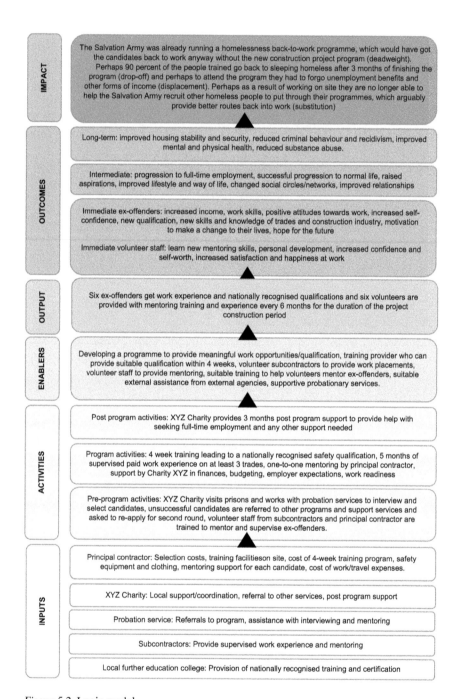

IMPACT

The Salvation Army was already running a homelessness back-to-work programme, which would have got the candidates back to work anyway without the new construction project program (deadweight). Perhaps 90 percent of the people trained go back to sleeping homeless after 3 months of finishing the program (drop-off) and perhaps to attend the program they had to forgo unemployment benefits and other forms of income (displacement). Perhaps as a result of working on site they are no longer able to help the Salvation Army recruit other homeless people to put through their programmes, which arguably provide better routes back into work (substitution)

OUTCOMES

Long-term: improved housing stability and security, reduced criminal behaviour and recidivism, improved mental and physical health, reduced substance abuse.

Intermediate: progression to full-time employment, successful progression to normal life, raised aspirations, improved lifestyle and way of life, changed social circles/networks, improved relationships

Immediate ex-offenders: increased income, work skills, positive attitudes towards work, increased self-confidence, new qualification, new skills and knowledge of trades and construction industry, motivation to make a change to their lives, hope for the future

Immediate volunteer staff: learn new mentoring skills, personal development, increased confidence and self-worth, increased satisfaction and happiness at work

OUTPUT

Six ex-offenders get work experience and nationally recognised qualifications and six volunteers are provided with mentoring training and experience every 6 months for the duration of the project construction period

ENABLERS

Developing a programme to provide meaningful work opportunities/qualification, training provider who can provide suitable qualification within 4 weeks, volunteer subcontractors to provide work placements, volunteer staff to provide mentoring, suitable training to help volunteers mentor ex-offenders, suitable external assistance from external agencies, supportive probationary services.

ACTIVITIES

Post program activities: XYZ Charity provides 3 months post program support to provide help with seeking full-time employment and any other support needed

Program activities: 4 week training leading to a nationally recognised safety qualification, 5 months of supervised paid work experience on at least 3 trades, one-to-one mentoring by principal contractor, support by Charity XYZ in finances, budgeting, employer expectations, work readiness

Pre-program activities: XYZ Charity visits prisons and works with probation services to interview and select candidates, unsuccessful candidates are referred to other programs and support services and asked to re-apply for second round, volunteer staff from subcontractors and principal contractor are trained to mentor and supervise ex-offenders.

INPUTS

Principal contractor: Selection costs, training facilitieson site, cost of 4-week training program, safety equipment and clothing, mentoring support for each candidate, cost of work/travel expenses.

XYZ Charity: Local support/coordination, referral to other services, post program support

Probation service: Referrals to program, assistance with interviewing and mentoring

Subcontractors: Provide supervised work experience and mentoring

Local further education college: Provision of nationally recognised training and certification

Figure 5.2 Logic model

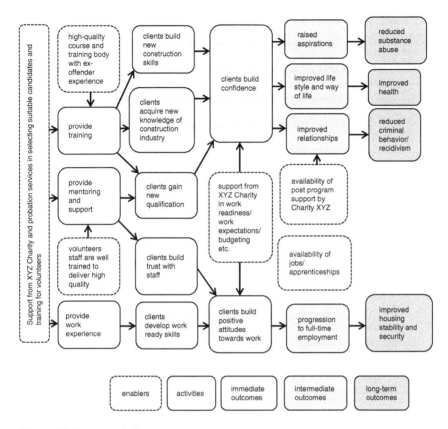

Figure 5.3 Outcome chain

can contribute to change for individuals, organisations and society, although one should not try to capture them all in a single diagram. The aim of an outcomes chain is to create a high-level depiction of the main causes and effects involved in bringing about change and highlight how the programme works by representing the critical elements that are fundamental to a project's success.

Narrative theory of change

It is also good practice to develop a narrative theory of change to add further depth and clarity to any visual representation you may choose to produce. A narrative theory of change can add particular detail around a programme's context, assumptions, evidence and enablers, which cannot be easily represented in a summarised visual map. The context element should describe the economic, social and political environment in which an organisation and its beneficiaries operate, who its beneficiaries are, and what change the programme is designed to bring about. You should also describe any assumptions you have made about what will happen in this environment upon which your intended change might depend, supported by any evidence on which these assumptions are based and any events/risks which might cause them not to happen. Evidence should also be provided that supports any cause and effect relationships you have identified, and the enablers that need to happen for your theory of change to happen should also be described. An example of a narrative theory of change for the ex-offender programme, described above, is provided below:

Narrative theory of change context:

- Much of our construction work occurs in disadvantaged communities where there is a high rate of unemployment.
- The construction industry has low barriers to entry for ex-offenders due to the nature of the skills required.
- Our programme provides workplace training and paid work experience for ex-offenders who want to seek work in the construction industry after prison.
- We work with a range of external organisations including training providers, charities and the probation service to select suitable candidates and provide a supportive and meaningful experience for them to gain skills and knowledge and experience to enable them to transition back to work and make a positive contribution to the community.
- There are numerous causes of offending behaviour such as long-term unemployment, past criminal records, family background, social circles, substance misuse, financial problems, lack of understanding of social norms, and lost hope and self-respect.

Assumptions:

- Providing employment and training in a closely supported and supervised workplace environment helps ex-offenders to transition back to work, by providing them with a sense of accomplishment, positivity, confidence, hope and direction after they leave prison.
- It also provides them with housing stability and security through an income for the first six months of leaving prison and no need to revert back to crime to earn an income.
- Work also changes the social circles that ex-offenders mix in and by doing so, develops new social relationships, role models and expectations.

Evidence:

- Ex-offenders who cannot gain employment or stable housing after leaving prison have been found to be more likely to reoffend.

- High levels of boredom resulting from unemployment after leaving prison have been identified as a significant factor that predicts violent reoffending.
- There is evidence that positive social networks are a relevant factor in adult reoffending.

Enablers:

- Developing a programme to provide meaningful work opportunities/ qualification; training provider who can provide suitable qualification within four weeks.
- Volunteer subcontractors to provide work placements.
- Volunteer staff to provide mentoring.
- Suitable training to help volunteers mentor ex-offenders.
- Suitable external assistance from external agencies.
- Supportive probationary services.

The general rules of theories of change

Whilst we have described the main forms a theory of change can take, there are no hard-and-fast rules. In essence, in does notmatter what a theory of change looks like as long as it has the following key attributes:

- Depicts the main elements of a programme.
- Shows the underlying rationale and causal logic of how programme inputs and activities lead to intended outcomes and impacts (including assumptions and enablers).
- Is easy to understand for a programme's key stakeholders (are all causal links visually clear and easy to follow?).
- Shows different chains for different stakeholder groups and depicts clearly the sequence of change that each stakeholder goes through when they engage in the programme.
- Includes both positive and negative outcomes.
- Has been developed through consensus between key stakeholders.
- Is supported by evidence.
- Is useful for designing and testing a programme that works.
- Is realistic, achievable and credible within capacity and resource constraints.
- Is verifiable in that all inputs, activities and outcomes are measureable (include quantities of inputs, outputs and outcomes where relevant).

Step 3 – Developing a measurement framework

This stage of the social value assessment process involves identifying and prioritising metrics for measuring whether the key material outcomes as identified by the theory of change have happened. It also involves establishing how they will be measured using valid and reliable methods of data collection, instruments and analysis.

The impact of your programme is likely to be complex, affecting different stakeholders in different ways, and over different time frames, which means that trying to capture all these changes will be challenging. The theory of change will be crucial in simplifying this process, since it will have identified the key material immediate, intermediate and long-term outcomes for each key stakeholder, which will need to be measured to determine the overall impact of the programme. However, a theory of change is only able to identify the 'expected' outcomes, so it is also important to be sensitive to unexpected outcomes that are likely to arise during the course of the programme. Furthermore, whilst the theory of change should have identified both positive and negative outcomes, there is clearly a danger in any assessment exercise of overly focusing on positive outcomes, especially if one has a stake in the design of the programme and in the outcomes being seen as positive by sponsors. This brings us to the critically important subject of how we minimise bias in social impact assessments. This is too often ignored in many impact assessments, making them unreliable and open to criticism.

Minimising bias in social impact assessments

Psychologists have revealed a large range of potential cognitive biases (both conscious and unconscious) which, if not controlled, can potentially introduce systematic error into the social impact assessment process. In simple terms, this bias can either arise from the behaviour of respondents providing data or from the behaviour of the researcher involved in its collection or interpretation. Below we list and describe most of the sources of bias, which will need to be controlled for in an effective impact assessment:

- Acquiescence bias – this is the tendency of respondents to agree with what the assessor says because they think the assessor is an expert (halo effect), and they are worried about the consequences of disagreeing. This may be because they are in an environment where they could be exposed for disagreeing, because they are friends with the assessor, from fatigue in answering long surveys or interviews, or because they are chosen from a sample which is inherently biased towards to assessor's the interests.
- Habituation –this is the tendency of respondents to fall into the habit of answering questions in the same way. This can arise because of the way questions are asked (respondents tend to provide the same answers to questions that are worded in similar ways) or because the sample comprises people who will tend to answer questions in the same way (for example, respondents who live in the vicinity of nuclear power stations get used to its presence and tend to downplay the risk).
- Framing bias – this is the tendency of assessors to prime respondents to respond in a particular way by the words and ideas presented in questions and

even the order in which questions are asked. To minimise this bias, it is crucial that assessors avoid asking leading questions and instead mix questions up to change the subject of the question and the types of answers that are likely to be provided.

- Anchoring bias – this is the tendency of assessors to rely too heavily on the first piece of information encountered when making an assessment of social value. This is why it is important to collect data from a range of stakeholders at different intervals during the intervention.
- Availability – this is the tendency of assessors to make judgements of value based on how easily or recently data can be brought to mind. This is why it is important that assessments conducted at the end of a project include information on impacts over its whole life cycle.
- Representativeness – this is the tendency of assessors to ignore long-term trends in data in favour of new data that might discount it. This is also called the 'base-rate error'. This is why it is important to take numerous measurements on social impacts throughout the life of a project and not rely on the last three measures.
- Sunk cost – this is the tendency of respondents to judge something's value based on the amount of time and resources they have invested in it.
- Affect – this is the tendency of assessors to let first impressions inform a decision, even if subsequent facts weigh against it.
- Fluency – this is the tendency of assessors to attribute greater value to data that can be processed more easily at the expense of those that are difficult to process.
- Recognition – this is the tendency of assessors to afford familiar recognised objects higher value than new objects.
- Similarity – this is the tendency of assessors to make judgments based on the similarity between current situations and past situations and that the circumstances and conditions underlying past valuations still hold true today.
- Social proof – this is the tendency of respondents or assessors in a group who are unsure of the correct way to value something, to look to others to do so.
- Peak-end rule – this is the tendency of respondents or assessors to judge the value of an event on how it ended rather than on how it went.
- Apophenia – this is the tendency of assessors to perceive meaningful patterns that do not actually exist within random data.
- Attribution – this is the tendency of respondents or assessors to attribute value based on certain respondent attributes: for example age, ethnicity and gender.
- Confirmation: this is the tendency of assessors to filter out information that does not confirm existing beliefs, especially when issues are emotionally charged – sometimes called selective perception/ cognitive dissonance.
- Culture – this is the tendency of respondents or assessors to see things from one's own cultural norms and perspective.
- Self-serving – the tendency for information to be distorted by the researcher to maintain their own interests.

- Status quo – this is the tendency to interpret information in a way which maintains the status quo.
- Stereotyping – this is the tendency to judge people or situations based on preconceived and ill-informed assumptions about that person or organisation.
- Social desirability – this is the tendency to interpret information in a way which conforms with existing social norms and societal expectations.
- Optimism bias – this is the tendency to interpret information in line with one's expectations – if expectations are better than reality, the bias is optimistic; if reality is better than expected, the bias is pessimistic.

In minimising these potential biases in social value assessments there are various well-established techniques one can use. Examples include using a team of assessors with different perspectives (insiders and outsiders), triangulating data using a mix of qualitative and quantitative methods to cross reference results, carefully sampling to ensure respondents reflect the target population making sure to collect a range of views both positive and negative and designing and asking questions in the right way, in the right forum, in the right location, and at the right time to elicit accurate and balanced responses. By thinking about these issues in advance, and designing the social value assessment process to minimise bias, the validity and reliability of the results is improved, giving credibility to any results generated.

Sampling

Effective sampling strategies are key to minimising bias when the potential numbers of people affected by a project or programme are too large to include everyone within the resource and time constraints of a social impact assessment. In this situation, a smaller group of people (the sample) have to be chosen to represent that population. It is critical in any social value assessment that the exact characteristics of the sample are stated, why they were selected and how they were selected to provide data. In simple terms the aim is to ensure that the people from whom data is collected accurately represent the population of stakeholders affected by your project or programme.

Sampling techniques can either be probability-based (random sampling) or non-probability-based (non-random sampling).

Probability-based (random) sampling ensures that every person in the population has an equal and unbiased chance of being chosen to provide data. This can be done in a number of ways. For example, simple random sampling, as the name suggests, involves randomly picking names out of a hat or numbers off a list manually or electronically. Systematic random sampling involves choosing a random starting point and then selecting the rest of the sample at equal intervals from that starting point. Stratified random sampling involves deliberately dividing the population into non-overlapping groups and then sampling from each of those groups. Cluster random sampling involves breaking the population randomly into groups and then randomly picking people out of all the groups.

Whilst probability-based (random) sampling can save time and expense, most researchers do not have the resources to randomly sample an entire population and therefore find it necessary to employ non-probability sampling. In non-probability sampling, respondents are selected on the basis of a purposive personal judgement of the researcher. This can be done in a number of ways. For example, convenience sampling involves picking your sample according to what is available. Snowball sampling involves a researcher asking one target person in the population to nominate other potential participants based on their knowledge so that the sample gradually snowballs to become larger. Judgement sampling involves a researcher deciding which population members to include based on his or her judgement, which is justified in some way. Quota sampling includes a designated number of people with certain specified characteristics. The downside of non-probability sampling is that the choice can be biased by the researcher's preferences and perceptions, and an unknown proportion of the entire population is not sampled, which means that the sample may or may not represent the entire population accurately and, as such, the results of the research cannot be generalised to the entire population.

Timing data collection

In terms of deciding the timing and frequency of data collection to minimise bias, practical considerations around resources will also need to be considered. There may also be sponsor, funder or other stakeholder requirements that require data to be collected at certain intervals and even from certain groups. In general terms, since social impact assessment involves measuring change as a result of a programme, one will need to collect data before (baseline data), during (programme data) and after (follow-up data) an intervention.

The timing of follow-up data collection after a programme depends on the time that changes are anticipated to last (the drop-off period), how long change takes to eventuate, the types of changes that one is interested in, the accessibility of data (respondents) after the programme has finished and practicalities such as time and resource constraints. However, follow-up measurements need to be treated with care and carefully justified, since it is easy to overstate your impact and lose your credibility by measuring outcomes for too long after an intervention has finished. As Nicholls et al (2012) notes, there is a difference between the life of an asset/capital project and its benefit period in the community.

Metrics, tools and indicators

In addition to ensuring the 'reliability' of an assessment process (the degree to which an assessment tool produces stable and consistent results) through considered sampling, one must also ensure it has high 'validity' (does it measure what it purports to measure?). Such validity is ensured by selecting the right indicators (metrics), and tools (measurement instruments), to measure social outcomes.

Since there is, as yet, no one commonly accepted standard for assessing social value, and since social programmes vary so widely in terms of the outcomes they produce and stakeholders they affect, there has been a proliferation of tools, methods and metrics for measuring social impact over the last decade. Indeed, the Social Value Portal in the UK, an online solution that helps organisations measure and manage the contribution that they make to society, estimates that there are over 1,150 social and environmental metrics being used for social impact assessment around the world. Many of these tools and metrics are sector-specific and impact-specific and vary widely in their scope, application, methods and cost. Take for example, self-esteem, mental wellbeing and physical wellbeing outcomes used in many social impact assessments. NPC (2014b, 2014a) notes that there are literally dozens of standard psychological and medically based tools designed to measure these outcomes, some designed as diagnostic tools rather than to measure changing outcomes and some based on questionnaires, whilst others centre on interviews, focus groups and even observations of beneficiaries in practice.

Clearly, when one starts claiming to have improved people's mental health (and any other specialist impact for that matter), there is the potential for construction professionals to face a minefield of potential criticism from experts in those areas (if they are not well informed). Indeed, there is a legitimate question about whether construction professionals should be attempting to measure social impact at all, since most are simply not qualified to measure social outcomes with any degree of confidence or credibility. Furthermore, while many off-the-shelf outcomes tools exist, which have been widely tested and validated, most still need expert interpretation to be understood. The often followed route for companies that operate in the construction sector, of paying expert researchers to measure social impact, is becoming increasingly expensive as the big consulting companies join the social impact bandwagon. On the other hand, while designing one's own tool might produce something suited to your stakeholders' needs, it is time consuming and to ensure credibility involves not just designing questionsbut the difficult process of testing them for validity, reliability and sensitivity.

To resolve these problems, some argue that the aim of impact measurement is not necessarily to produce an academically precise measure of impact, but instead to tell the story of impact from the point of view of the respondents, rather than report some objective facts using scientific method. Nevertheless, whether one agrees with it or not, there is an inexoriable trend towards making social impact assessment a science through the use of objective, rather than subjective, methods and indicators. So to overcome potential criticisms and add credibility, validity and reliability to the way that social value is measured, it is useful to follow the basic principles of social impact measurement outlined at the start of this chapter which in simple terms requires that an impact assessment:

- prioritises what it seeks to measure on material outcomes for key stakeholders (rather than trying to measure every single change that occurs in the population);

- is done in consultation with key stakeholders;
- is supported by evidence collected through research that is both valid, reliable and verifiable;
- is undertaken within time and resource limitations;
- is undertaken in a transparent manner;
- is easy to communicate;
- is compliant with legislative and industry frameworks;
- covers all outcomes that are relevant to your programme and stakeholders' businesses;
- is balanced and does not over-claim.

Since there are so many different metrics that can be used to measure social value across so many impact areas, it is also useful to be aware of the increasing number of emerging social value banks, data sets, standards, tools and frameworks for measuring social value. These resources, which vary from very simple to highly complex, allow practitioners of social value measurement to exchange their methods of measurement and experiences and develop globally accepted and standardised protocols within and across industries to measure social value (although none yet exists in the built environment). For example, at the simplest level, and specifically designed for the construction industry, is a simple checklist of social value indicators provided by the Considerate Contractors Scheme for its signatories. At the end of a project, a Considerate Contractors Scheme monitor visits a site, collects the data and produces a report based on the information collected, which a contractor can use to show it has generated social impact to its stakeholders.

A far more sophisticated approach called the Supply Chain Social Value Bank (SVB) has been developed by Morgan Sindall and Daniel Fujiwara (who developed the wellbeing approach to valuation discussed in more detail below) with the support of Scottish Water. The SVB monetises and reports on a suite of metrics that align to the activities within the construction industry that create a value to society and which are measurable. The purpose of the SVB is to imbed social value priorities into construction activity from the outset of a project rather than bolting it onto mainstream activities. It is also designed to identify where latent social value sits, capture it and drive more efficient processes to deliver enhanced value to society from existing activity. The benefit of the SVB to a contractor is that it reports on individual project, business unit and company-wide performance. It also allows for benchmarking based on construction type and postcode, and analysis of the data held will facilitate the allocation of resource to activities that create the greatest value to society.

Another recent development of interest to those in the construction sector is the National Themes, Outcomes and Measures (TOMs) Framework for Measuring Social Value (2018) published by the National Social Value Taskforce (2018). These TOMs support the implementation of the Public Services (Social Value) Act 2012 by proving a clear definition of social value and a corresponding measurement framework that provides transparent and robust social value measurement, reporting and comparison.

The framework produces a standard set of five social impact themes, 18 outcomes and 35 measures/indicators and 'proxy values'. It also provides a social impact calculator that allows procuring bodies to compare tenders in a way that is proportional and relevant to the bid and able to better justify a procurement decision.

The five themes and 18 outcomes are as follows:

Theme 1 – Promoting Skills and Employment: To promote growth and development opportunities for all within a community and to ensure that they have access to opportunities to develop new skills and gain meaningful employment.

Outcomes:

- more local people in employment
- more opportunities for disadvantaged people
- improved skills for local people
- improved employability of young people.

Theme 2 – Supporting the Growth of Responsible Regional Businesses: To provide local businesses with the skills to compete and the opportunity to work as part of public sector and big business supply chains.

Outcomes:

- more opportunities for SMEs and VCSEs
- improving staff wellbeing
- a workforce and culture that reflect the diversity of the local
- community
- ethical procurement is promoted
- social value embedded in the supply chain.

Theme 3 – Creating Healthier, Safer and More Resilient Communities: To build stronger and deeper relationships with the voluntary and social enterprise sectors whilst continuing to engage and empower citizens.

Outcomes:

- crime is reduced
- creating a healthier community
- vulnerable people are helped to live independently
- more working with the community.

Theme 4 – Protecting and Improving our Environment: To ensure the places where people live and work are cleaner and greener, to promote sustainable procurement and secure the long-term future of our planet.

Outcomes:

- climate impacts are reduced
- air pollution is reduced
- better places to live
- sustainable procurement is promoted.

Theme 5 – Promoting Social Innovation: To promote new ideas and find innovative solutions to old problems.

The taskforce also recognises that the national TOMs are designed primarily to reflect the needs of different industries. In response, and over the coming months, the Social Value Taskforce will be publishing sector 'plug-ins' that reflect the specific needs or opportunities of individual sectors such as construction. Nevertheless, this framework is likely to be too complex for the vast majority of construction businesses who are still struggling to understand what the term social value means. Furthermore, whilst the use of a standard checklist approach has obvious advantages, it will inevitably limit people's thinking to focus on these types of outputs at a macro level and to look at value in financial terms even though the guidelines recognise that this is not the case.

There are of course many hundreds of other generic social value banks and data bases and calculators that can be used. For example, TRASI (Tools and Resources for Assessing Social impact) is a database of philanthropists, foundations and grants with links to over 200 tools developed by these organisations, andmeasure social impact. IRIS is a library of common indicators and metrics (built on over 40 sector-specific standards and reporting frameworks) for organisations in different industries to choose from. Created by the Global Impact Investing Network in 2009 (a non-profit organisation dedicated to increasing the scale and effectiveness of social impact investing), IRIS is used by over 5,000 organisations and is designed to produce a common language for social impact investors to describe an organisation's non-financial performance. The Global Value Exchange is another free open source database of social values, outcomes, indicators and stakeholders run by Social Value UK. This is designed to allow social impact information to be shared, creating greater consistency and transparency in measuring social and environmental values. Staying in the UK, the New Economy Unit Cost Database was established in 2009 by the Greater Manchester Local Enterprise Partnership to support social cost benefit analysis of economic development and public service reform projects and programmes. Based on national costs derived from government reports and academic studies, this unit cost database brings together more than 600 cost estimates of social outcomes in areas relating to crime reduction, education and skills provision, employment provision, health outcomes, housing and social services.

Some social value banks are sector-specific, such as the Housing Association's Charitable Trust (HACT) Value Bank, which contains over 600 social indicators and values in the areas of employment, local environment, health, financial inclusion and youth. HACT is a social enterprise-based ideas and innovation agency for the UK Housing Association sector and launched their Social Value Bank in 2014, making it freely available for housing associations and on licensing terms for organisations operating in other sectors. HACT also provides its own online social impact calculator to help housing associations assess and measure the social impact of repairs, maintenance and neighbourhood regeneration projects on the lives of tenants who live in their properties. The HACT approach is based on a wellgine approach to valuation developed by Daniel Fujiwara (Fujiwara 2013) that measures the success of a social intervention by how much it increases people's wellbeing, based on large UK national surveys isolating the effect of a particular factor on a person's wellbeing. Analysis then reveals the equivalent amount of money needed to increase someone's wellbeing by the same amount. The main argued advantage of the wellbeing valuation approach, which is used by a range of leading UK authorities and government departments, is that it represents a consistent, robust and comparable method for placing values on things that do not have a market value through being bought and sold. It is argued that this is because these values are based on data from people's actual experiences self-reported wellbeing and life circumstances, which is in contrast to other valuation methods that are based on how people perceive their life, introducing potential psychological complexities and biases. The values are calculated through statistical analyses of four large national UK datasets that contain data on people's responses to wellbeing questions and questions on a large number of aspects and circumstances of their lives such as employment status, marital status, health status, whether they volunteer, whether they play sports, whether they live in a safe area and so on, resulting in a wide range of values. Since many government departments in the UK use this approach, this allows robust comparison across different types of intervention.

The UK National Health Service (NHS) Outcomes Framework is a set of 68 indicators that measure performance in the health and care system at a national level. Indicators in the NHS Outcomes Framework are grouped into five domains that set out the high-level national outcomes that organisations in the NHS should be aiming to improve. These include: preventing people from dying prematurely; enhancing quality of life for people with long-term conditions; helping people to recover from episodes of ill health or following injury; ensuring that people have a positive experience of care and treating and caring for people in a safe environment and protecting them from avoidable harm. Each year since the NHS Outcomes Framework was first published in 2010, the Department of Health has been improving the framework by refining existing indicators and developing new indicators.

At the next level down, there are other tools and frameworks that are programme and cohort-specific. For example the Journey to Employment (JTE) Framework, of potential value to social value programmes that focus on employment outcomes, identifies seven groups of social impact indicators for the potential of a young person securing and sustaining a job. These indicators include:

- Personal circumstances (access to resources such as transport or the internet, risky behaviours such as alcohol or drug problems, and family issues such as caring responsibilities).

- Emotional capabilities (self-esteem, having grit and determination to succeed).
- Attitudes (general feelings about participating in work and their aspirations); employability skills (communication, teamwork and leadership skills).
- Qualifications, education and training (qualifications and attainment, as well as conduct and behaviour).
- Experience and involvement (work experience, involvement in the community, and networks developed as a result).
- Career management skills (having career direction, an interest in enterprise, understanding how to search for jobs and present themselves to employers).

The Outcomes Star produced by a social enterprise called Triangle Consulting is a family of evidence-based tools in the form of a simple star diagram, which presents a set of key dimensions of change on a scale of 1–5 that are relevant to various impact areas and sectors. It allows users to track change in social value indicators over time across a range of sectors including: adult care, armed forces veterans, autism and ADHD, community, criminal justice, domestic violence, education, employment, families and children, health, housing and homelessness, substance abuse and young people. There are now over 30 versions of the star, each developed through research and consultation with leading bodies in their sectors. They are available under licence and training and used widely in the UK by national and local charities, local authorities, the NHS, police, schools, housing associations, care and support services. The outcomes star is also used by a growing number of organisations across Europe, Asia, Australasia and the USA and by 2016, over 60 organisations had collaborated on or funded the development of versions of the star. It has been translated into ten languages, including Danish, French, Italian, Dutch and Chinese, with more in the pipeline. For example, the Youth Outcomes Star captures social change in six areas of young people's lives: making a difference; hopes and dreams; wellbeing; education and work; communicating and choices and behaviour. The Mental Health Outcomes Star is designed for adults managing their mental health and recovering from mental illness and covers ten key areas: managing mental health; physical health and self-care; living skills; social networks; work; relationships; addictive behaviour; responsibilities; identity and self-esteem and trust and hope.

Another type of tool available for measuring social impact is the Local Multiplier (LM3) online tool that enables an organisation to calculate the local economic multiplier effect on its community. LM3 takes its name from the Keynesian multiplier, which has been used since the early twentieth century to measure how income entering an economy circulates within it. Developed by UK social entrepreneur Adam Wilkinson, in partnership with Northumberland County Council and the New Economics Foundation, LM3 automates the calculation process working from an upload of spending data, such as a contract or company turnover and a target local area. In this way, it allows companies, government or community organisations to measure how their spending generates local and non-local economic impact and benefit to the community in which they operate. In simple terms, LM3 measures the multiplier effect of income into a local economy over three rounds of spending, taking into account an organisation's turnover or project cost including procurement and employee wages and other forms of cost; where and with whom the company spends that money and where and how suppliers and employees re-spend their incomes. The multiplier is then calculated for every unit of currency spent within a local area selected by the user. For example, an LM3 multiplier of 1.50 would indicate that for every £1 created by a social programme, an additional £0.50 is generated for the local area. The LM3 website claims that LM3 has now been applied to over £13 billion of spending in public, private and not-for-profit sectors, with many large- and medium-sized enterprises using the tool as an auditable, transparent, objective and directly comparable way to measure, demonstrate and maximise the social value of their CSR activities, business activities and supply chains on local economies.

Most recently, the Social Return on Investment Network launched Wiki VOIS, a database of values, outcomes and indicators. The database is suitable for any organisation interested in the outcomes of its work, and is designed to develop more commonality in the use of social value outcomes, indicators and values in undertaking SROIs. It also provides a resource for analysts and evaluators in all sectors and an international forum for people to share, discuss and develop information on social value outcomes, indicators and values.

Given the above, it is not surprising that there is so much confusion around the many different metrics and methods that organisations can use to measure and report their social impact. While some organisations simply report anecdotes and stories or the results of customer satisfaction surveys about the benefits of their products and services, others invest enormous amounts of time and resources in using sophisticated measurement techniques such as SROI which require significant expertise, research and empirical data collected before, during and after an intervention and in many cases, long-term follow-ups and randomised control trials to form the basis for any conclusions drawn. Perhaps unsurprisingly, those that invest the most time and resources tend to be larger organisations with significant resources.

What is happening in the built environment?

There are as yet no social value banks, data sets, standards, tools and frameworks for measuring social value dedicated to the built environment. However, a number of bespoke privately funded frameworks are being developed in the construction industry by firms such as Wates, Wilmot Dixon and Interserve plc working in association with leading universities and other industry bodies such as the Construction Youth Trust and Constructing Equality. Nevertheless, the vast majority of the construction industry are well behind these pioneers and are not even aware of the concept of social value, let alone thinking about the integration of such measures into their performance measurement and reporting systems. Clearly, the built environment has a long way to go in meeting the increasing requirements to measure and report its social impact. As discussed in Chapter 1 and above, there are many challenges in building understanding of social value assessment in the built environment from such a low base. However, there is a great opportunity to learn from other sectors where the proliferation of measurement methodologies is being resolved by the establishment of common principles of measurement that are appropriate to those sectors. For example, Inspiring Impact (an alliance of eight representative organisations from across the third sector) has developed a blueprint for shared measurement based on the premise that similar types of organisations addressing similar social issues should develop shared measurement protocols suited to their common needs. We are not proposing that a common approach to social value assessment be established across the entire built environment. Given the wide variety of businesses, projects and communities in which the industry builds, it is unlikely that one single approach would be appropriate to all situations. However, common principles of measurement could be established within project or company supply chains, within different sectors of the industry (for example housing, infrastructure and commercial) or at different organisational scales (large, medium- or small-sized businesses). This would have a number of important advantages such as: greater relevance of measurement indicators to the activities, goals and types of organiations in specific businesses, projects and supply chains; improved communication around social value because a common language can be used; savings in time and resources; greater comparability, tracking and benchmarking to facilitate learning and improvement; better collaboration between organisations which are seeking the same goals; improved reliability and quality of impact measurement over time; and clarifying an organisation's impact on a particular social problem. To enable this to happen, NCP (2014a) argues that leadership needs to be shown by a controlling organisation, to bring different organisations together in a coordinate approach, in order to collaboratively develop, negotiate and define a shared understanding of what social value means and what outcomes are desired, including how to measure and report them reliably and practically to enable comparison and learning to occur.

Data collection and analysis

Having decided on one's theory of change, objectives in terms of material outcomes and the metrics and the tools for assessing them, the next step is to develop a data collection and analysis strategy. Every metric and tool requires data and the best social impact assessments use a combination of quantitative and qualitative techniques of data collection and analysis using both primary and secondary data collected in a variety of ways and from a variety of stakeholders and sources. This is known as triangulation or methodological pluralism and allows an analyst to cross check their findings in various ways giving greater credibility to the results being presented (Bryman 2012).

In simple terms, quantitative techniques are based on the methods used in the natural sciences (biology, chemistry, physics, geology) and are underpinned by an assumption that there is only one universal truth/reality about the social value and this can be measured and quantified in some meaningful way. So the aim of quantitative methods is to 'measure' and 'count' social value in quantifiable terms collecting numerical data based on questions about how much, how many, how often and to what extent.

In contrast to quantitative methods, qualitative methods are based on approaches to data collection and analysis developed in the humanities and social sciences and are based on a belief that there is no single ultimate truth about social value and there are multiple interpretations of social value from a range of different perspectives that need to be taken into account in any assessment of social value.

Primary data is new raw data collected by the investigator from people affected by a project or programme using techniques such as surveys, interviews, focus groups and observation. In contrast, secondary data is data that already exists from previously published research such as official statistics, media reports, diaries, letters and websites, although the context and quality of the research should always be checked and verified before it is used.

Ethics in social impact assessment

As a final note in this section, it is important to appreciate the importance of ethics in undertaking your social impact assessment. This is too often missing from social impact assessment literature but critically important for a whole range of reasons, not least because social impacts assessment often involves collecting data from vulnerable groups and the research process itself has the potential to do more harm than good if not conducted in a sensitive and responsible manner. According to Shamoo and Resnik (2015) ethical research should adhere to the following basic ethical principles:

- honesty (does not fabricate, falsify, or misrepresent data);
- objectivity (avoids bias and discloses personal or financial interests that may affect research);
- integrity (keep promises and agreements; act with sincerity);
- carefulness (avoids careless errors and negligence; keeps good records of data, to enable scrutiny by others);
- openness (share data, results, ideas, tools, resources);
- respect for intellectual property (honors patents, copyrights, and other forms of intellectual property and gives proper acknowledgement or credit for all contributions to research);
- confidentiality (protect the anonymity of respondents and confidential communications and data);
- responsible publication (publish in order to advance research and scholarship and duplicative publication);
- respect (respect your respondents and colleagues and treat them fairly);
- social responsibility (prevents harm to the community and to respondents and colleagues);
- non-discrimination (avoids discrimination against colleagues or respondents or the community on the basis of sex, race, ethnicity etc.);

- competence (informed by good research practice, professional competence and expertise);
- legality (complies with relevant laws and institutional and governmental policies);
- care (shows proper respect and care for people and animals);
- protection (minimise harms and risks and maximise benefits to human respondents/subjects and respects human dignity, privacy, and autonomy; taking special precautions with vulnerable populations; and striving to distribute the benefits and burdens of research fairly).

Documenting the measurement framework

After considering all of the above, the assessment framework can be represented in a simple table like Table 5.1, listing the outcomes, metrics and tools selected to measure them, and how the data will be collected and analysed to identify the social impact of the social programme.

Step 4 – Implementing the programme and assessing and monitoring social impact

Having established community needs and priorities, created a programme and its theory of change and developed a reliable and valid measurement framework to collect and analyse social outcome data, it is now time to implement the social programme and assess its social impact.

Programme implementation is often missing from literature on measuring social impact, yet it is a crucial stage of the social value creation process. Implementation is typically much more difficult than developing strategy and no

Table 5.1 An example of a measurement framework

Short-term, intermediate- and long-term outcomes	Metric	Research design
Improved mental health	Mental Health Index (insert reference to chosen index)	Anonymous on-line survey of youth candidates who have gone through the programme undertaken in week 1 (start) and week 12 (end) of the programme.
Improved housing stability	Housing stability index (insert reference to chosen index)	Interviews with youth 24 weeks after end of programme around 8 main dimensions: housing type, recent housing history, current housing tenure, financial status, standing in the legal system, education and employment status, harmful substance use, and subjective assessments of housing satisfaction and stability.

matter how well thought through your theory of change and programme are, poor implementation can undermine any intended social impacts, making the process of assessment futile.

Stages in the implementation process

The implementation of a social programme doesn't happen in one step and to help us understand the process in more detail, Rogers (1995) proposes a widelyrespected five-stage model:

1. knowledge stage;
2. persuasion stage;
3. decision-making stage;
4. implementation stage;
5. confirmation stage.

The *knowledge stage* is the stage when stakeholders are told about the social programme's existence and aims to gain some understanding of how it works and the benefits for them. Key stakeholders will need to be provided with information about a programme and its merits before they are likely to engage with it. For some stakeholders, where there is a recognised need for the programme before they engage with it (demand-led social innovation), receptivity to information will be high. In contrast, for others where there is no recognised requirement to engage with the social programme, information must be presented in a persuasive way that creates a need that was not previously understood (supply-push innovation). Sinek (2009) argues that persuading potential stakeholders to adopt a new idea is a complex process and so the *persuasion stage* can take considerable effort and time to get to the 'tipping point' where stakeholders will start engaging with the programme on mass. Sinek's work indicates that the fastest way to reach this point is to focus your message on the reason for the programme (the 'why') rather than what the programme entails (the 'what') and to aim this message towards stakeholders with similar values and beliefs, thereby maximising the chances of securing an emotional connection to your idea. Social reinforcement is also an important part of the persuasion process because research tells us that people make decisions, not only on the basis of their own opinions, but also on the opinions of others with whom they are socially connected (Lundvall 2010). So it is at the persuasion stage of the social programme implementation process that 'change-agents' and 'opinion leaders' play their greatest roles in overcoming system 'norms', which often act as a barrier to the adoption of a new idea. The *decision stage* involves stakeholders making a decision about whether to engage with the programme or not. This stage is often the most difficult and as Rogers (1995) points out, most new initiatives get adopted at a frustratingly slow pace. Once accepted by stakeholders the social programme can be implemented in order to accomplish the social outcomes stated in the theory of change. It is at the *confirmation stage* that stakeholders seek feedback about the effectiveness of the social programme that has been put in place. They get this information from official sources (for example the protagonist designer of the programme) and from unofficial sources (their own business and other stakeholders involved in the programme). Based on this information, they will decide to continue supporting the programme or to walk away. Therefore, the ability to measure and report programme success is critical to the sustainability of any social programme and it is to these issues that we turn in the next sections.

Assessing social impact

The process of assessing social impact involves assessing whether the planned outcomes in your theory of change are actually achieved in practice (and any other unplanned unexpected outcomes which you did not predict). In other words, whether there is a material change (both positive and negative) in the lives of the stakeholders concerned, net of counterfactuals such as: deadweight (what would have happened anyway); drop-off (reducing benefit over time); attribution (what else could have contributed to the change); displacement (what other benefits does the intervention displace/push aside) and substitution (replacement of other gains).

GECES (2014) notes that there are a number of risks which need to be managed during this stage.

- Not being focused – It is important to use your theory of change to focus your measurement on the material outcomes that stakeholders said were important to them. Do not try to measure everything.
- Not maintaining proportionality – The basic principles of social impact measurement require that impact measurement needs to be proportionate to the programme and resources available.
- Not being transparent – The basic principles of social impact measurement require that there is underpinning evidence to support your claims and that the process needs to involve key stakeholders and be transparent to them.
- Missing important outcomes – Measuring change in outcomes is unlikely to be something that can be done at the end of the intervention. First, there is a need for baseline data to be collected at the start of the programme to enable change to be measured. Second, change occurs over the life of a programme in terms of short-term, intermediate- and long-term outcomes, which means that data can be lost if one waits to the end. Finally, there will be unexpected outcomes that occur in unpredictable ways and times that will need to be captured as they arise. So the collection of impact data should be a continuous process to be undertaken over the life of the delivery of a social intervention.
- Over-claiming – The golden rule in counting value is that a benefit/outcome can only be counted once. For example, if a long-term outcome of your intervention was someone getting a job, one could legitimately measure the tax benefits to the government from the increased employment. Nevertheless, when counting the monetary benefits of increased income to the person who

got the job, one would have to deduct the tax that goes to the government since this has already been counted. To avoid over-claiming, data collection should also take account of the drop-off of outcomes over time and not continue beyond a point when claims about change become excessive and indefensible.

- Outputs being treated as outcomes – it is a common mistake in social impact assessment to count output (such as training 30 people) as outcomes. It is important to remember that while outputs such as training numbers are easy to measure, they are only a means to achieving the ultimate objective of a social programme which is to bring about affective, cognitive, behavioural and situational changes in people's lives (the outcomes).

- Quantification at the expense of understanding – It is widely accepted that although there have been some advances in quantitative approaches to social impact assessment (such as SROI) in recent years, there can be a tendency to over-rely on numbers as an indicator of social impact. However, as discussed earlier, numbers can only provide a limited insight into the rich and nuanced reality of how a social programme changes people'slives. The narratives and stories of how change happens in people's lives should always be a central part of any social impact assessment.

Valuing outcomes

One of the most contentious questions in measuring impact is whether it is appropriate to evaluate social impact in monetary terms.

For example, advocates of cost benefit analysis (CBA), an established technique used widely by organisations to assess policy outcomes, argue that it is possible to place a monetary value on the social, economic and environmental costs and benefits associated with projects and programmes. CBA used discount rates to convert future costs and benefits back to present day net present values. In addition, probabilistic and sensitivity analyses are used to judge how each cost/benefit ratio changes under different circumstances. However, critics of CBA point to the inherent difficulties in assigning financial values to the many intangible outcomes produced by an organisation and argue that they will tend to be overlooked or undervalued. Furthermore, CBA focuses on one perspective (the policy maker's), thereby ignoring the views of other external stakeholders who may experience change as a result of a social programme.

For this reason, other techniques such as cost-effectiveness analysis (CEA) are sometimes used where it is inappropriate to monetize effects and where factors other than costs are important in making decisions. For example, in the realm of health and safety, CEA recognises that placing an economic value on someone's life is problematic both ethically and practically. Other measures of outcome are therefore developed such as 'years of life gained' (in quality adjusted life years). Similarly, in comparing the social impact of different social procurement programmes such as those which create employment opportunities for disadvantaged youth, competing options would be compared not only for their monetary benefits and costs and savings, but also for such factors as the impact on crime, community resilience and health.

Despite long-standing concerns over monetisation of social value, an increasingly popular approach to assessing social value called Social Return on Investment (SROI), does exactly this (Vik 2016). According to CM (2014), SROI is also the main method used in the construction sector to measure social impact, although it should be noted that only a few major companies are experimenting with its use with variable success and, compared to other industries, there are precious few examples of SROI published in construction (see for example, Bridgeman et al. 2015, 2016; Kelly and Whitley 2016).

In simple terms, SROI methodologies seek to capture the wider social, economic and environmental benefits of an organisation and translate them into a monetary value and single all-encompassing SROI ratio. Emerging out of the traditions of CBE and social accounting and drawing on the principles and practices of social impact measurement, SROI avoids the limitations of CBA by acknowledging the need to measure value from multiple stakeholder perspectives. The growing popularity of SROI has been driven largely by the strong support of influential bodies such as the UK and Scottish Government, The SROI Network, New Economics Foundation, Social Value International, New Philanthropy Capital and various academics (Social Value UK 2010, Brouwers, Prins and Salverda 2010, Nicholls et al. 2012).

In essence, SROI monetises all the inputs and material outcomes (short-term, intermediate- and long-term material outcomes) for key stakeholders associated with a social programme, discounted to present day values, and takes into account various counterfactuals such as deadweight, drop-off, attribution, displacement, and substitution. Nicholls et al. (2012: 9) outline six key steps in their widely used guideline for those wanting to use SROI to assess their projects and programmes:

1. Establish scope and identify key stakeholders – define boundaries of who will be involved in the process and how.
2. Mapping outcomes – develop a theory of change in consultation with key stakeholders.
3. Evidence inputs and outcomes and give them a value – collect data around input and outcomes (both negative and positive) and value them in monetary terms.
4. Establish impact – estimate and deduct counterfactuals.
5. Calculate SROI – compare inputs to outputs and test for sensitivity.
6. Report, use and embed the results – share results with key stakeholders and use feedback to improve the impact of the programme in the future.

For example, casting our minds back to the example we used to explain a theory of change, by attaching a monetary value to the improvement in an ex-offender's mental health following employment on a construction project, an SROI practitioner may decide to include in their valuation the extra income the job generated for that person, which is easily computed. In addition, the calculation could include the cost savings of not attending mental health clinics, of not having to buy mental health drugs, of not having to pay fines and the forgone wages for having to take time off work. The monetary values used to make these calculations are called 'financial proxies' in recognition that they can only ever be an approximation of the true value of a change.

Using a standard spreadsheet, SROI calculates the total social impact by adding these up (over a defined measurement period – deducting deadweight, attribution, displacement and substitution) and discounting this net value back to current day values to create a single Net Present Value (NPV) figure. Drop-off is used to measure the reducing benefit over time for any outcomes that last for more than one year and is usually expressed as a percentage. So for example, if an outcome such as increased income from a job (valued at $10,000 in year 1) lasts for three years and the drop off per year is 50 percent because the hours reduce by 50 percent per year, then the total outcome value after counterfactuals have been deducted (the social impact) is $10,000 year 1 + $5000 year 2 ($10,000 – 50 percent) + $2500 year 3 ($5000 – 50 percent) = $17,500. The final step of SROI is the calculation of the SROI ratio which is the total NPV of the impact divided by the total NPV value of the inputs. The ratio indicates how much social impact is generated for every $ invested. So for a ratio of 1:1.50, every $1 invested in the programme produces $1.50 of value.

In the case of 'forecast' SROIs (that look forward and anticipate impacts as opposed to 'evaluative' SROIs that look backward at actual impacts) it is also good practice to undertake a risk/opportunity analysis and sensitivity analysis by challenging the assumptions underlying the outcomes, quantities and proxies and any counterfactuals that form the basis of the SROI calculations. By recalculating the impact and SROI ratio using a range of best case and worse case scenarios, it is possible to produce a distribution (possibly with probabilities attached) of SROIs across that range of scenarios that reflect the risk involved and also the potential variability in the SROI calculation. Such a process will also help to reveal which factors and assumptions are critical to the outcomes required and the amount and type of change that could turn the impact from a positive to a negative. For example, if a sensitivity analysis found that the social impact of a programme is especially dependent on one particular charity being involved, then one would divert resources to managing this risk. Finally, it is also possible to calculate a payback period by dividing the inputs by the annual impact, since there may be investors involved who need to know this information.

Understandably, the process of reducing everything down to a monetary value is highly attractive to many policy-makers and managers. Money is a common language and currency everyone in business and government understands. However, the monetisation principle underpinning SROI has also attracted significant criticism. As Nicholls et al. (2012) note, although markets have developed as mechanisms to determine the monetary value of something, the attribution of monetary value to social outcomes, that often do not have a market price, must invariably involve the use of approximations (proxies) of monetary value by drawing on data that was intended for other purposes and collected in different contexts. For example, attaching a monetary value to someone's improvement in mental health (assuming this can be measured anyway) by using the health service cost of treating someone for mental health is inherently flawed, since this is generally an average cost based on a sample across multiple types of mental health for a cross section of patients in multiple geographical areas and times. This inevitably means that the output of an SROI is inherently subjective and involves a large number of assumptions which are often not revealed in the data that is being used. For example, it may not be correct to assume that the increased self-confidence of getting a job is likely to mean much more to someone from a disadvantaged background who has been unemployed for five years than it is to a graduate from a privileged background who has just left university?

Critics of SROI argue that the crucial difference between a tangible product or service which has a market value attached and a 'soft' social outcome such as improved mental health which does not, is that there is no agreed price or monetary value that one can reliably attached to it. However, in SROI's defense, Nicholls et al. (2012) argues that no product or service has any *actual* inherent objectivevalue other than what people in a market agree. All that markets do is provide a behavioural mechanism and place which allows a range of people to come together and agree what that price should be. Therefore, it is wrong to assume that just because a social outcome does not have a market (and therefore an agreed price) it does not mean that it has no monetary 'value' to people. It merely means that the market has not been established to agree this price and that this market can be artificially created by a good social impact assessor using a variety of mechanisms by asking a range of stakeholders what they think something is worth – a technique which has a long history in environmental and health economics. For example, there are various techniques such as stated preference/ contingent valuation methods such as The Value Game which are used in SROI to produce these monetary values. Essentially, most of these involve asking stakeholders to value the outcomes they experience relative to other products they also value (which have a market price), or how much they would pay, or compensation they would accept, to avoid something. So, for example you may ask someone what their happiness is worth in terms of other products in a product catalogue or what compensation they would accept for the noise created by a new road. However, as Nicholls et al. (2012) point out, there are no standard methods of doing this and although social value banks are starting to develop to allow SROI practitioners to standardise the social outcome valuation process, the reality is that most evaluations of social value are highly variable and underpinned by little rigorous science and research and are therefore easily criticised for having low validity. It then follows that since there is such high variability in the way that data is collected and the proxies used to measure value in SROI studies that the eventual SROI ratio is of little worth and certainly cannot be used for comparison between different programmes that have been carried out by different assessors.

As we discussed earlier, these criticisms become even easier when the range of outcomes being measured grows into areas such as mental health where specialist expertise is needed to understand the concepts and phenomenon being measured.

Problems can also arise when SROI practitioners do not understand the way that the proxies they are using are constructed, since many are not credible. Consider, for example, calculating the monetary benefit of giving an ex-offender a job on a construction project by calculating the savings from the reduced recidivism this has produced (itself a questionable assumption). As Nicholls et al. (2012) point out, if this has been calculated by simply dividing the total cost of prison service by the number of prisoners then this will overestimate the social value since a prison has both fixed (which remain the same when one prisoner is removed) and variable costs (which change when one prisoner is removed). Strictly, it is only the variable costs which should be claimed. However, this type of cost information is rarely available to outsiders.

Finally, in calculating social impact, there are also potential complexities and pitfalls in estimating for counterfactuals. For example, to calculate deadweight one needs to make comparisons to control groups and other similar benchmarks as representative of your population as is possible. So if one is measuring the impact on re-offending rates of employing ex-offenders on a construction project, then one would need to compare re-offending rates for people on the project with national or regional re-offending rates. Furthermore, in terms of displacement, it is difficult to know if someone had to give up another job to take the training opportunity on a construction project and, in terms of displacement, it is difficult to estimate whether a policy to employ ex-offenders may be denying employment opportunities to other needy groups in the community. Finally, in estimating attribution, it is difficult to consider what other factors people and/ or organisations might be contributing to the change you are claiming to have brought about. For example, if you are claiming to have caused someone's self-confidence to have increased, you will need to consider the possibility that this may have been caused by them attending other programmes. Collecting this type of data is likely to be well beyond the resources and expertise of a construction contractor.

In addition to the above criticisms of subjectivity, unreliable data, inconsistent methodologies, inbuilt biases and dubious assumption, numerous other criticisms of SROI have been put forward including the expense and time involved in undertaking the process or employing a consultant to do this and gaining external certification. There are also those who have a fundamental philosophical opposition to monetising social value and who question the large gap between theory and practice in most assessments. Finally, critics point out that the outputs of SROI cannot be compared or relied upon due to lack of standardised approach, varying resources and time invested and who undertakes the process. The reality is that few people have been trained to do an SROI in the construction industry (and outside); internationally or even nationally recognised qualifications do not yet exist and skill levels therefore vary considerably.

The best defence for much of this criticism is to use official statistics and valid and reliable methods as much as possible. Practitioners of SROI also claim that the process does not pretend to produce an accurate measure of benefit and that if the measures of value are evaluated in consultation with stakeholders and make sense to them, then that is all that ultimately matters. For example, in construction, Murdock and Bridgeman (2014: 13) argue that the output of an SROI should 'not be restricted to one number, but seen as a framework for exploring social impact, in which monetisation plays an important role but not an exclusive factor'. Furthermore, as Arvidson et al. (2010) point out, transparency in conducting an SROI is essential, with any assumptions, judgments and sources of data used being well documented and open to scrutiny. Independent external verification of the results is also useful.

Step 5 – Reporting to stakeholders, learning and improving

As the social programme is implemented and social impact assessments emerge at regular intervals, results need to be reported regularly and meaningfully to internal and external stakeholders and discussed at a collaborative level to enable the programme to continually improve. It is important to do this throughout the implementation process, rather than wait until the end, to avoid missing key lessons to improve the programme. In simple terms, the aim of your reporting is to combine the qualitative and quantitative impact data that has been collected and analysed in order to tell the story of change in an unbiased and balanced way. In this way it needs to take into account the perspectives of the stakeholders in the most concise and clear way possible. A good report should also acknowledge any assumptions that have been made in arriving at the results, any possible counterfactuals and any limitations that may qualify the results.

Some important questions that need to be considered in developing an effective approach to reporting social value to stakeholders include:

- Who are the target audiences?
- What information does each want out of the reporting process, in what form and how will they use it?
- What are our objectives in communicating with them?
- What mediums are best suited to each target group?
- What is the best communication style and how does this align with cultures and values?
- How do we give our performance reports credibility and avoid being accused of rhetoric?

Meaningful social impact reports should follow the guidance we have provided in this chapter and at each reporting point should inform a discussion about how the programme might be adapted to ensure, and ideally improve, its effectiveness in achieving intended outcomes into the future.

Finally, and ideally, although resources may not permit, some external independent endorsement is useful in providing credibility and independence, especially when there are many stakeholders involved with conflicting interests. GECES (2014) suggests three levels of endorsement are appropriate in different circumstances:

1. Validation – This is part of the normal research process and involves providing appropriate supporting evidence for all social outcome assessments and claims.
2. Independent review – This requires an independent party to review the measurement process and findings and comment upon their completeness and the underlying logic.
3. Audit assurance – This is a more formalised approach requiring the issuing of a pre-worded assurance certificate that the report is a 'true and fair' reflection of the social impact produced by the programme. For full SROIs this involves commissioning organisations such as Social Value International. However, since the methods and standards of reporting are so variable and yet to be standardised, there are no organisations which accredit non-SROI reports at present.

In terms of reporting the social impacts of programmes in annual accounts at 'whole-of-organisation level', it is notable that, although sustainability reporting remains largely voluntary in end-of-year company reports in most countries, reports of social impact are increasingly incorporated into company annual reports, either as separate sustainability reports or as part of an integrated annual report covering triple bottom line economic, social and environmental indicators. For example, in Australia, the new Australian Stock Exchange Corporate Governance Principles and Recommendations (ASX 2014), which came into force in July 2014, require all publicly listed companies to disclose exposure to economic, environmental and social sustainability risks. The board of directors has a responsibility to ensure any report is accurate, unbiased, transparent and complies with the law (and in the case of public companies, any stock exchange listing requirements). Failure to comply with these basic expectations can potentially lead to legal action, de-listing and, at the very least, reputational damage among the increasing numbers of external stakeholders who monitor business performance. When organisations have under-performed, reporting in this way is a brave thing to do and there is a temptation for many executives to hide the bad news or move the goalposts to make the outcomes appear more palatable. However, this is unwise given the sophistication and degree of scrutiny now applied; there is nothing more damaging than being accused of distorting the facts.

In the absence of globally accepted standards for reporting social performance, many organisations, both large and small, refer to auditable standards, guidelines, tools and indicators developed by globally respected bodies such as the United Nations, World Business Council for Sustainable development (WBCSD), Social Accountability International (SAI), International Organization for Standardisation (ISO) and the International Integrated Reporting Committee (IIRC). While these have been developed for large organisations, some of the principles are relevant to small and medium-sized businesses and are outlined below.

- United Nations Global Reporting Initiative (GRI) produces a range of indicators to use for reporting around environmental impacts, labour practices, human rights, social and product responsibility (including a construction and real estate supplement with a number of additional indicators of building energy and water usage, carbon emissions, asset remediation, OHS, displacement/resettlement of people and sustainability certification). The GRI is the world's most widely used standard for sustainability reporting.

- SA8000 is produced by Social Accountability International (SAI) and is a certification standard that is designed to streamline industry and corporate codes, and to create a common language and standard for measuring social compliance. It can be applied to any company in any industry and provides indicators of performance in eight areas important to social accountability in the workplace. These include: child labour; forced labour; health and safety; free association and collective bargaining; discrimination; disciplinary practices; working hours and compensation.

- AA1000 is produced by AccountAbility, an independent, global, non-profit organisation promoting accountable, responsible and sustainable business practices and corporate responsibility. The AA1000 standards require all disclosures to comply with three main principles of effective reporting: materiality, completeness and responsiveness.

- The International Standards Organization (ISO) is a non-governmental standards organisation made up of 163 member country standards organisations. It is the world's largest developer of voluntary international standards and guidelines, designed to assure consumers of products and services that certified products conform to the minimum standards set internationally. ISO 26000:2010 'Social Responsibility' has been issued to contribute to global sustainable development by encouraging business and other organisations to practice social responsibility by improving their impacts on their workers, their natural environments and their communities. As discussed in Chapter 1, ISO 26000 provides guidance rather than requirements and encourages organisations to design and report their business activities with the following areas in mind: organisational governance; human rights; labor practices; environmental impacts; fair operating practices; consumer issues; community involvement and development.

- The International Integrated Reporting Committee (IIRC) was established in 2011 and aims to create a globally accepted International Integrated Reporting Framework that enables organisations to provide performance, strategic and governance information in a clear, concise and comparable format reflecting the commercial, social and environmental context in which they operate. After a number of discussion papers and draft consultation reports, the IIRC released its first International Integrated Reporting Framework in December 2013 (IIRC 2013) and it defines an integrated report as:

a concise communication about how an organization's strategy, governance, performance and prospects, in the context of its external environment, lead to the creation of value over the short, medium and long term.

(IIRC 2013: 7)

However, Glass's (2012: 96) research into sustainability reporting in the construction sector found that integrated reporting is currently 'a tall order for the construction sector' because at the time only one of the 64 global companies that use it are from construction.

Summary

The assessment of social impact is clearly a highly complex and inherently subjective (and some would say inaccurate) process. Ultimately, it should be remembered that the process of assessing social impact is not about the measurement itself but about driving behaviours amongst stakeholders that can maximise positive outcomes for the disadvantaged in the community. It is crucial to use the information produced not only as a reporting mechanism to stakeholders for funding, statutory or reputational reasons, but as a way to improve the programme. The danger in all assessment systems is that people game the system and focus on the targets themselves forgetting the ultimate objective of the assessment process was to benefit specific groups. As with all assessment, it is also important to ensure it does not drive perverse behaviours, which may, perhaps unsurprisingly, be counter to these stated aims. For example, in an attempt to achieve targets, there can be a tendency for stakeholders to focus on beneficiaries that take less effort and resources to manage, at the expense of the difficult cases who really need help.

The field and practice of social impact assessment is in its early stages and, in the future, will need to develop more standardised measurement protocols and conventions driven by research to understand and advance impact assessment best practice. Whilst the built environment is not a special case, there are certain attributes requiring an industry-specific lens, and this needs to be driven by commercial, third-sector and community organisations working together to continuously improve impact assessment practices and learn from one another. Such a process will need to be supported by socially conscious clients, investors, researchers and organisations able and willing to act as champions and custodians of standardised impact assessment methods, data and metrics. Moreover, they need to provide thought leadership and forums to enable market actors to share experiences and work with them to improve measurement over time including increasing transparency and accountability for impact delivered.

Key points

- Social impact assessment as the process by which those in the built environment account for all types of value that their businesses, projects or programmes create, beyond just economic impacts to include social, environmental, cultural and health impacts at both a community and individual level.
- It is not possible to quantify and monetise all forms of social value and it is for this reason that we have decided to use the term 'assessment' rather than 'measurement'.

- Social impact assessment remains one of the most contested issues in policy, research and practice. Research in this area is still in its infancy and while many good reports, reviews, best practice guidelines and toolkits have been produced, there remains a lack of conceptual clarity about social impact measurement with very few robust, comprehensive and empirical studies to draw from.
- Being able to assess social value is crucial for a number of reasons: it enables more informed decision-making about how to maximise social impact; it is crucial in communicating effectively with key stakeholders; it allows managers to incentivise positive behaviour by linking social value outcomes to rewards and sanctions.
- The assessment of social value/impact can be predictive and forward looking (a forecast impact assessment) or retrospective (an evaluative impact assessment).
- To assess the social value of a social initiative/programme accurately one must be involved throughout its life (inception, through design, implementation and close-out), since understanding the goals of a programme are crucial to understanding its impacts; many of which may take time to eventuate.
- It is important at all times to adhere to the seven core principles of social impact assessment: involve stakeholders; understand what changes; value the things that matter; only include what is material; do not over-claim; be transparent and verify the result.

References

AccountAbility (2008) *AA1000 AccountAbility Principles Standard 2008*. AcountAbility UK Ltd: London.

Achleitner, A., Bassen, A., and Roder, B. (2009) An integrative framework for reporting of social entrepreneurship. *Social Science Research Network (SSRN), working paper series, paper 1325700*. Available online at: http://papers.ssrn.com/sol3/papers.cfm?abstract_id=1325700 [accessed 27th November, 2014].

Airhihenbuwa, C.O. (1995) *Health and Culture Beyond the Western Paradigm*. Sage Publications: Thousand Oaks, CA.

Altschuld, J.W. (2010) *The Needs Assessment KIT*. Sage Publications: Thousand Oaks, CA.

Arnstein, S.R. (1969) A ladder of citizen participation. *Journal of the American Institute of Planners*, 35(4): 216–224.

Arvidson M., Lyon F., McKay, S., and Moro, D. (2010) The ambitions and challenges of SROI. *Third sector research centre, working paper 49, December*. University of Birmingham: Birmingham, UK.

ASX (2014) *Corporate Governance Principles and Recommendations, 3rd edn*. ASX Corporate Governance Council: Sydney, Australia.

Barnes, P. (2002) Approaches to community safety: risk perception and social meaning. *Australian Journal of Emergency Management*, 15(3): 15–23.

Baumol, W.J. (2010) *The Microtheory of Innovative Entrepeneurship*. Princeton University Press: Princeton, NJ.

Brewer, J.F. (2013) From experiential knowledge to public participation: social learning at the community fisheries round table. *Environmental Management*, 52(2): 321–334.

Bridgeman, J., Murdock, A., Maple, P., Townley, C., and Graham, J. (2015) Putting a value on young people's journey into construction: introducing SROI at Construction Youth Trust. In: Raiden, A. and Aboagye-Nimo, E. (eds), *Proceedings 31st Annual ARCOM Conference, 7–9 September 2015.* Lincoln, UK, pp. 207–216.

Brouwers J., Prins, E., and Salverda M. (2010) *Social Return on Investment: A Practical Guide for the Development Cooperation Sector.* Cornelis Houtmanstraat: Utrecht, The Netherlands.

Bryman, A. (2012) *Social Research Methods*, 4th edn. Oxford University Press: Oxford.

Burby, R.J. (2001) Involving citizens in hazard mitigation planning: making the right choices. *Australian Journal of Emergency Management*, Spring: 45–58.

Burns, D., Hambleton, R., and Hoggett, P. (1994) *The Politics of Decentralisation: Revitalising Local Democracy.* Macmillan: London.

Chinyio, E. and Olomolaiye, P. (2010) *Construction Stakeholder Management.* Wiley-Blackwell: Oxford.

Clandinin, D.J. and Connelly, F.M. (2000) *Narrative Inquiry: Experience and story in qualitative research.* Jossey-Bass: San Francisco, CA.

Cleland, D. and Ireland, L. (2007) *Project Management: Strategic Design and Implementation*, 5th edn. McGrath Hill: New York.

Close, R., and Loosemore, M. (2015) Breaking down the site hoardings: attitudes and approaches to community consultation during construction. *Construction Management and Economics*, 32(7–8): 816–828.

CM (2014) Social value: gearing up for giving back. *Magazine of the Chartered Institute of Building.* 3 April, available online at: www.construction-manager.co.uk/agenda/social-value-gearinggiving- back [accessed 3rd July, 2014].

CSI (2014) *The Compass: Your Guide to Social Impact Measurement.* Centre for Social Impact: Sydney, Australia.

Davidson, E.J. (2005) *Evaluation Methodology Basics: The Nuts and Bolds of Sound Evaluation.* Sage Publications: Thousand Oaks, CA.

Department of Planning and Community Development (2011) *Procurement for Social and Economic Development Outcomes in Local Communities: Mapping and Analysis Methodology.* Department of Planning and Community Development: Victoria, Australia.

Fawcett S.B., Paine-Andrews, A., Francisco, V.T., Schultz, J.A., Richter, K.P., Lewis, R.K., Williams, E.L., Harris, K.J., Berkley, J.Y., Fisher, J.L., and Lopez, C.M. (1995) Using empowerment theory in collaborative partnership for community health and development. *American Journal of Community Psychology*, 23(5): 677–697.

Fujiwara, D. (2013) A general method for valuing non-market goods using wellbeing data: three stage wellbeing valuation. *Centre for Economic Performance Discussion Paper No. 1233.* Centre for Economic Performance, London School of Economics and Political Science: London.

G8 (2014) *Measuring Impact: Guidelines for Good Impact Practice. Report to the Working Group of the Social Impact Investment Taskforce established by the G8.* Available online at: https://thegiin.org/assets/documents/Webinar%20Slides/guidelines-for-good-impact-practice.pdf.

GECES (2014) *Proposed Approaches to Social Impact Measurement in European Commission Legislation and in Practice Relation to: EuSEFs and the EaSi.* GECES Sub Group on Impact Measurement: Paris.

Glass, J. (2012) The state of sustainability reporting in the construction sector. *Smart and Sustainable Built Environment*, 1(1): 87–104.

ICAEW (2015) *Quantifying Natural and Social Capital: Guidelines on Valuing the Invaluable*. Institute of Chartered Accountants in England and Wales: London.

IIRC (2013) *International Framework*. International Integrated Reporting Committee (IIRC): London.

International Organization for Standardization (ISO) (2010) *ISO 26000 – Social Responsibility*. International Organisation for Standardisation: Geneva, Switzerland.

Jones P., Comfort D., and Hillier D. (2006) Corporate social responsibility and the UK construction industry. *Preliminary Paper*. Available online at: https://hrcak.srce.hr/3755.

Littau, P. (2015) *Managing Stakeholders in Megaprojects*. University of Leeds: Leeds, UK.

Loosemore, M. and Phua, F. (2011) *Socially Responsible Strategy: Doing the Right Thing?* Routledge: London.

Lundvall, B.A. (2010) *National Systems of Innovation: Toward a Theory of Innovation and interactive learning*. Arnold: London.

Mitchell, R.K., Agle, B.R., and Wood, D.J. (1997) Toward a theory of stakeholder identification and salience: defining the principle of who and what really counts. *Academy of Management Review*, 22: 853–886.

Moodley, K. (1999) *Project performance enhancement – improving relations with community stakeholders, Profitable Partnering in Construction procurement*. Conference proceedings, Ogunlana, S.O. (Ed.). E&FN Spon: London.

Murdock, A. and Bridgeman, J. (2014) The construction youth trust: developing measure of social impact. Paper presented at the *20th Voluntary Sector and Volunteering Research Conference, 10th – 11th September 2014, Sheffield Hallam University, Sheffield, UK*.

National Resource Center (2010) *Conducting a Community Assessment. Strengthening Non-profits: A Capacity Builder's Resource Library*. Compassion Capital Fund National Resource Center. Available online at: http://strengtheningnonprofits.org/resources/guidebooks/Community_Assessment.pdf.

National Social Value Taskforce (2018) *National TOMs Framework 2018 for Social Value Measurement: Guidance*. National Social Value Taskforce: London.

NCVO (2013) *Inspiring Impact: The Code of Good Impact Practice*. National Council for Voluntary Organisations. Available online at: www.inspiringimpact.org/wp-content/uploads/2018/01/Code-of-Good-Impact-Practice-1.pdf.

Nicholls, J., Lawlor, E., Neitzert, E., and Goodspeed, T. (2012) *A Guide to Social Return on Investment*. Social Value UK: Liverpool, UK.

NPC (2014a) *Building Your Measurement Framework: NPC's Four Pillar Approach*. New Philanthropy Capital: London.

NPC (2014b) *Using Off-the-Shelf Tools to Measure Change*. New Philanthropy Capital: London.

NSW (2015) *Principles for Social Impact Investment Proposals to the NSW Government*. Office of Social Impact Investment, NSW Government: Sydney, Australia.

O'Connor, L. (2003) *A Socio-Technical Analysis of the Standard Based Apprenticeship in Ireland: A Case Study of the Construction Industry. Unpublished Thesis*: University of Leicester.

Our Community (2017) *Conducting a Community Needs Assessment*. Our Community Ltd: Melbourne, Australia.

Preece C., Moodley K., and Smith A. (1998) *Corporate Communications in Construction*. Blackwell Science: Oxford.

Pritchard, D., Ní Ógáin, E., and Lumley, T. (2013) *Making an Impact*. National Philanthropy Foundation: London.

Raiden, A.B., Dainty, A.R.J., and Neale, R.H. (2006) Balancing employee needs, project requirements and organisational priorities in team deployment. *Construction Management and Economics*, 24(8): 883–895.

Rogers, E.M. (1995) *Diffusion of Innovations*, 4th edn. The Free Press: New York.

Rotary (2015) *Community Assessment Tools: A Resource for Rotary Projects*. Rotary: Evanston, IL.

Shamoo, A. and Resnik D. (2015) *Responsible Conduct of Research*, 3rd edn. Oxford University Press: New York.

Sinek, S. (2009) *Start with Why: How Great Leaders Inspire Action*. Penguin: New York.

Social Value International (2016) *Understanding and Use of the Principles of Social Value*. Social Value International: London.

Social Value UK (2010) *Starting Out on Social Return on Investment*. Social Value UK: London.

Spence, W.R. (1994) *Innovation*. Chapman and Hall: London.

Tam, C.M. and Tong, T.K.L. (2011) Conflict analysis study for public engagement programme in infrastructure planning. *Built Environment and Project Asset Management*, 1(1): 45–60.

Teo, M.M.M. (2009) An investigation of community based protest movement continuity against construction project. *Unpublished Phd Thesis:* University of NSW: Sydney, Australia.

Tritter, J. and McCallum, A. (2006) The snakes and ladders of user involvement: moving beyond Arnstein. *Health Policy*, 76(2): 156–168.

Troast, P. (2011) Using web and social media in your business. *Home Energy Journal*, 28(6): 40.

Van Oosterhout, J.P.H. and Kaptein, M. (2008) The internal morality of contracting: advancing the contractualist endeavour in business ethics. *Academy of Management Review*, 3(3): 521–539.

Vik, P. (2016) What's so social about social return on investment? A critique of quantitative social accounting approaches drawing on experiences of international microfinance. *Social and Environmental Accountability Journal*, 37(1): 6–17.

Watkins, R., West Meiers, M., and Visser, Y.L. (2012) *A Guide to Assessing Needs: Tools for Collecting Information, Making Decisions, and Achieving Development Results*. World Bank: Washington, DC.

Watson, K.J. and Whitley, T. (2016) Applying Social Return on Investment (SROI) to the built environment. *Building Research & Information*, 45(8): 875–891.

Wilcox, D. (1999) *A to Z of Participation*. Joseph Rowntree Foundation: London.

Winch, M., Mucha, M., Roberts, M., and Shinn, B. (2007) *Sustainable Public Involvement – The Case for Lasting, Meaningful Stakeholder Partnerships*. Summer Edition. Washington State Public Works: Washington, DC.

Winch, G.M. (2004) Managing project stakeholders. In: Morris, P.W.G. and Pinto J.K. (eds), *The Wiley Guide to Managing Projects*. Wiley: Hoboken, NJ, pp. 321–339.

Witkin, B.R. and Altschuld, J.W. (1995) *Planning and Conducting Needs Assessments: A Practical Guide*. Sage: Thousand Oaks, CA.

Part 2

Case studies

Showcase of good practice

Part 2 of the book includes a selection of case studies to showcase good practice in creating and delivering social value in design and placemaking, building and construction, multiparty collaborations and housing. Given the plethora of toolkits and material available about practice in the public sector, we highlight case examples from the private and third sectors. Our aim is to encourage reflection and inspire action by showing what is possible when the legal, moral and business cases for social value come together and create impact.

6 Building social value into design and placemaking

Sophia de Sousa, The Glass-House Community Led Design

This chapter explores the connection between people and place and how the quality and qualities of the built environment around us affects our social, cultural and economic interactions. It unpacks the role that placemaking can play in improving people's quality of life, and its potential for social value creation through both the built outcomes and the design process itself.

Touching on some of the key parameters of placemaking and the different design practices that fall within it, the chapter will draw on case studies from The Glass-House Community Led Design, a national charity that supports communities, organisations and networks to work collaboratively on the design of buildings, open spaces, homes and neighbourhoods. These examples draw together different approaches to creating social value through placemaking, a fundamental way of creating positive change.

People and place

We all have a profound connection with the physical spaces and places around us. Our everyday places, where we live, work and play, have an enormous impact on our quality of life. They affect our health and wellbeing, and play a significant role in the social, cultural and economic opportunities that are available to us. They influence our interaction with others and help shape the people we become.

The complexity of the spaces and places around us and how we interact with them, is extraordinary, made more so by the diversity and complexity of any population. Every place contains and caters for many different people and their diverse needs, aspirations, and simply, their personal preferences.

We also have a huge impact on the places around us. We shape places through their design and construction but also through our everyday actions. How we manage and maintain spaces influences how they are used and enjoyed by people. How we behave in any given space affects the quality of that space and other people's use of it both at that moment and over time.

In this context, placemaking is a bold endeavour. The decision to transform a place, to design with the objective of considering and responding to the needs and aspirations of a complex local population, carries with it enormous responsibility.

Placemaking also offers great opportunity in terms of creating social value.

If we take for granted that the primary objective of placemaking is to improve the quality of a place, also implicit in that objective must be to improve the quality of life for the people who will live, work and play there. Social value, therefore, is clearly embedded in the objectives for placemaking. There is an expectation that the built outcomes of placemaking will have an impact on the place and on people long after the construction ends. Increasingly, we are measuring the impact and social value of placemaking in these terms and learning about how changes to our built environment affect people's social, cultural and economic opportunities and their overall quality of life.

However, the potential for social value in placemaking extends far beyond its built outcomes and their impact. The process of change itself offers a broad spectrum of opportunities to enhance the social value of any placemaking scheme. This chapter will explore specifically how The Glass-House Community Led Design (hereafter The Glass-House) engages local people in the design process in placemaking and creates some of those opportunities.

Empowering design practices at The Glass-House Community Led Design

The Glass-House is a national charity that supports communities, organisations and networks to work collaboratively on the design of buildings, open spaces, homes and neighbourhoods. It sees design not only as a tool for creating great places but also as a way to connect people and to empower them with enhanced confidence, skills, and a greater sense of agency. The Glass-House calls design processes that build in benefit for those who engage with them *empowering design practices*. In social value terms, this approach places a specific focus on what engaging in the design process itself can offer people and groups.

Empowering design practices begin with defining design in broad terms: from identifying local needs and aspirations to understanding the local context in physical and socio-economic terms through developing a shared vision and design brief to an inclusive, engaged and transparent design process. Such an approach can bring better informed physical changes, and perhaps changes in use, in the spaces and places in question.

Design is an iterative process, a journey, which moves through a series of stages, but it is rarely a simple, linear process. Each design journey is different, responding to its own context, its challenges and opportunities, and it draws together a complex combination of skills and knowledge, including the design professional, the client, stakeholders, user groups and so on. Engaging people in that process can happen at many different levels and at different stages in the journey.

Participatory design

One of the approaches that The Glass-House advocates and enables is a participatory design process in which local people and organisations are actively engaged in the design process being led by the commissioning client/project lead.

Participatory design goes beyond consultation to actively engaging local people and organisations in informing the vision and brief, in ideas generation and in decision-making.

Participatory design can happen at any scale and be led by any sector. It has enormous potential to build empowerment and social value in placemaking.

Setting out a participatory design process

If done well, participatory design can provide a clear and transparent process to transform a place. This starts with being able to answer some key questions. The Glass-House always asks:

1. Why should people get involved?
   How is this particular design/placemaking process relevant to them, and how might changes that occur as a result of the process benefit/affect them?

2. How will involvement affect the decision-making process?
   People need to know what difference their contribution will make to decision-making and where the opportunities for influence lie. Setting out clear project parameters, outlining which elements of the project are already fixed and where participation can help inform and shape decisions is crucial. Ensuring that participation is not just tokenistic but genuinely helping to inform the vision, brief and iterative design journey is essential.

3. How can people get involved?
   People need to understand the mechanics of getting involved in a design process. The range of opportunities for interaction might vary depending on the context, scale of the project, commissioning client or local interest. However, in every case, there is value in a clear strategy that sets out a design process and the means for supporting the iterative dialogue around it.
   Are the meetings and other engagement activities open to anyone who wants to join in? Will formal steering or advisory groups be convened, and if so, how does someone join such a group? Are there specific criteria for who can take part in any of the engagement activities? Will there be single or multiple opportunities to express opinions, to contribute ideas, to step into a creative co-design space?
   It is important to recognise that people will be able, and inclined, to engage in different ways and with different levels of commitment. As such, it is essential to create opportunities for people of different ages, cultural backgrounds experiences and abilities to contribute in ways that are both meaningful and empowering.

4. What can people contribute to the process?
   When we move beyond basic consultation in placemaking, the engagement of local people in the design process can unlock valuable assets for the design and placemaking process. Here are just a few:

- *User experience and knowledge*
  Whether this is of the local area in which a project sits, or of how a particular type of place (such as a youth centre, school or hospital) works, experiential knowledge is an invaluable resource to a design team.
- *Relationships and connections*
  Connections with other local people and organisations can help extend the reach and richness of the engagement process and can also help build the networks and partnerships that will support the success of the place in the long term.
- *Investment*
  Getting involved in shaping a place makes people more willing to invest in its success. This investment might be the time and energy to contribute to the dialogue, a contribution to the assets required to bring a project to fruition or to enliven the place once transformed.
- *Complementary projects/activities that add value*
  Engaging with local people may unearth other initiatives in the area that are complementary to a scheme or project. Working together can make the process for change more efficient and enhance the impact and legacy of both large and small projects in the area.
- *Creativity*
  There is great opportunity in opening up ideas generation and creative thinking. While this has to be managed through a clear process, it can be the source of locally driven innovation that responds to issues and opportunities in new ways that are borne out of, and responsive to, local context.

5. What will they take from the process?
   Being given the agency to take part in a design process and being able to participate in local decision making is hugely empowering, and its value should not be underestimated. Sometimes, however, it is not just the act of inviting people into the dialogue but also equipping them to contribute confidently that creates opportunities for social value and impact.

   Creating opportunities to build confidence, skills and relationships through training, enabling and other activities will help unlock people's capacity to contribute to the design process. It can also build valuable social capital through the development of life skills, employability and local networks: real social value.

   Every participatory design process should be able to answer the question of 'What's in it for them?. This key question is considered in terms of the social value of giving people the agency to contribute in a meaningful way to decision making and in terms of the personal, organisational and community development opportunities that the design process offers.

When considering the development of projects at a grassroots level, The Glass-House focuses on community-led and co-design processes, which may provide either complementary or integrated approaches within the frame of participatory design.

Community-led design projects within the built environment are led and commissioned by a community-based organisation or group. They include appropriating and using small pieces of disused space, revitalising or developing community buildings, open space projects and larger scale development through development trusts, community land trusts and community-led housing. Sometimes, a small action, such as community gardening and food-growing on a housing estate, can have an enormous impact by improving the place quality of a derelict area and in social value terms, for example, teaching people new skills, bringing people together andimproving health and wellbeing.

Larger scale projects offer an alternative to top-down development and build local investment into the process of development and regeneration and the long-term strategy for the use, management and maintenance of a place by local people.

When these locally driven actions and placemaking initiatives are rooted in local knowledge and engage a broad spectrum of local people in the design process, they benefit from a firm understanding of what already exists in an area and the gaps to be addressed. If they invite local people into decision making and actions and if they build on local conversations and partnerships, they stand a better chance of their impact and social value spreading beyond those who initiated the project in the first place. An inclusive and participatory community-led project can thus extend its reach and the social value it produces.

Collaborative/co-design takes place when various groups and interests come together to commission and/or develop a design process that responds to both their individual and collective needs and aspirations. What sets co-design apart from the participatory design process described above is that it is conceived as a more democratic space in which all partners share in equal measure both the agency and responsibility of setting the agenda, building a vision, defining the brief and taking the project through the various stages of design. Co-design is born out of collaboration and partnership, whereas participatory design invites it in.

In many ways co-design in truly democratic terms is an ideal and is perhaps challenging to implement in large-scale projects. However, it is both interesting and possible to build co-design principles and opportunities, which may take the form of specific elements or initiatives within a larger scheme into any participatory design process.

Design as a vehicle for social value in placemaking

Empowering design practices within placemaking can embed social value at a number of different levels. This includes:

- Building public awareness, interest and investment in the built environment.
- Improving the quality of places.
- Increasing access to quality places – improving place equality.
- Empowering people to play an active role in shaping their environments (including the possibility of influencing policies and practice).

- Creating opportunities for capacity building and skills development through a participatory design process (i.e. design skills but also team building, creative arts practices, communication, life skills, employability, etc).
- Taking people outside of their own environment to be exposed to different places and approaches and to be inspired.
- Building social networks and helping people to be more active and empowered citizens.
- Unearthing and mobilising assets through collaboration.

What follows is a series of stories from The Glass-House practice that illustrate how these aspects of social value can emerge through participatory design processes.

White City Green

Working with a development team from St James, a member of The Berkeley Group, and with Phoenix Academy, formerly Phoenix High School (near White City in Shepherd's Bush), The Glass-House developed a programme for 30 students in Year 10 that aimed to help local young people engage with the design process for a large, open space within the emerging White City Green development (The Glass-House 2014a). In addition to the design focused benefits, from the outset all partners saw this programme as an opportunity to also help the young people develop their confidence, skills and employability.

The starting point was to get young people exploring their immediate surroundings and also further afield. They were invited to articulate their experience of the places around them and to talk about their interactions with others in the public spaces in their neighbourhood. This offered local expertise to the

Figure 6.1 Students on a study visit[1]

Figure 6.2 Students presenting their design ideas

development team, providing a better understanding of the local context and issues as experienced by young people.

For the young participants, it also offered an important space for reflection on the environment around them and on the many interactions that occur in the public realm. When asked to step outside their own experiences to explore the needs, concerns and aspirations of others, the young people were no longer just contributing to a design process but also formulating a better understanding of their place in society and developing their empathy for others. Participation in this built environment project offered a frame in which the young people could explore important societal issues like citizenship, diversity and disability.

The workshops also offered the young people a view into the development industry and the employment opportunities it offers. Working with the staff at St James, the young people were introduced to the broad spectrum of roles within the sector. For many of the young people, who had grown up in a disadvantaged area with high rates of unemployment, it was the first time they had a sense of opportunity for employment in their future. Two of the students went on to do work placements at St James.

Through a series of workshops that saw them developing and articulating their views and design ideas, the young people also developed valuable confidence and skills. Alongside the design training and creative activities, such as drawing and model making, which were crucial to helping the young people to engage constructively in the dialogue around the emerging scheme design, they also worked on their communication and presentations skills. This allowed them to present their ideas to the St James Board of Directors articulately and with confidence.

The young people also shared their experience and ideas with a number of local community organisations and businesses at an event that brought together local stakeholders. This was an opportunity for the young people to connect with local networks and groups, and to be seen differently, as valuable contributors to local initiatives.

Feedback from students

> My experience in the White City project was unquestionably fascinating. It has helped me improve my skills, in particular communication and confidence. It made me realise that people want to hear what I have to say.
>
> (Programme participant)

Key opportunities for social value:

- Building public awareness and interest in the built environment.
- Empowering people to play an active role in shaping their built environment.
- Building the confidence of young people as active citizens, exposing them to different people, experiences and environments and to different ways of thinking.
- Building confidence, skills for life and employability through a participatory design process.
- Building social networks.

Tidworth Mums: a case for soft play

Light-touch design engagement activities built on collaboration across sectors can add huge value to an area. When the Tidworth Mums, a volunteer-run group dedicated to championing and providing creative play activities in Tidworth, Wiltshire set out to explore the potential for a soft play centre, their engagement activities helped bring their local community and different sectors together to imagine the future of play in their area (The Glass-House 2014b).

Working with Tidworth Mums and partners The Open University, Wiltshire Council and the Army Welfare Service, The Glass-House helped design and deliver a Mega Soft Play Day, in order to talk to local people about play provision in the area. The day aimed to assess local appetite for a new soft play centre and to work with children and families to explore design ideas. The Mega Soft Play Day was designed as a low-cost, high impact event, and drew on the collective skills, experience and assets of the partners involved. This collaboration formed part of an interdisciplinary and cross-sector action research project called Unearth Hidden Assets through Co-design and Co-production (The Glass-House 2014c), which was led by Brunel University and was funded by the Arts and Humanities Research Council through their Connected Communities Programme.

The Mega Soft Play Day was initially conceived by Tidworth Mums as a low-cost play activity for children and their families during half term. The Army Welfare Service and Wiltshire Council, which did not at that time offer local play services of their own, had been working with and supporting Tidworth Mums and

Figure 6.3 Members of the research team talking to children about where and how they like to play at the Mega Soft Play Day and supporting them to illustrate their ideas

saw this as an opportunity to work collaboratively to enhance the provision of services for local families. They offered the group the use of the main hall at their jointly owned and managed leisure centre, as well as the staff time of members of their local teams.

The research team saw an opportunity in this event to explore local views on play, and to engage with local people on how they would view and might use a potential new soft play centre. The day also offered a fantastic opportunity to talk to local children and their parents and carers about the type of play they enjoyed and to help them advise on the design of potential new play facilities and events. Using simple and playful research methodologies within a pop-up soft play environment, the team were able to gather data, views and ideas to inform next steps. At the same time, the Mega Soft Play Day provided a vital and affordable local play service to 275 children and 158 parents/carers, in an environment in which children, their siblings and their parents/carers could play together.

One of the key findings from the day was that whilst there was clearly an interest in a soft play centre, most of the benefits could be achieved by mobilising existing assets collaboratively. Through partnership across sectors modelled on the Mega Soft Play Day, soft play activities could be provided and kept affordable and accessible to local people. There have since been many more Mega Soft Play Days.

Figure 6.4 Families playing and testing equipment at the Mega Soft Play Day

Another important outcome of the collaboration between Tidworth Mums, Wiltshire Council and the Army Welfare Service was that Tidworth Mums was recognised by local people, partners, and the group itself as a valuable asset within the community. They became local champions for play and indeed helped inform policies within the council and Army Welfare service regarding play provision.

The transient nature of the army community, with some members of the group moving to new locations and new members joining over time, has meant that learning from these activities spreads to new people and locations. The Glass-House and the Open University also produced a resource, *Tidworth Mums: A Case for Soft Play* (The Glass-House 2014b), to help capture the story and learning from this project and make it accessible to other groups trying to improve play provision and facilities in their areas.

Feedback from a participant

> Being involved in the research has given me new insights into what is meant by 'assets'. We used to see these as physical things, such as buildings, but we now recognise that people are assets, and that our group, the Tidworth Mums, is a valuable asset for the community.

(Research participant)

Key opportunities for social value:

- Empowering people to play an active role in shaping their built environment.
- Mobilising skills and assets through collaboration across sectors to shape places (be it temporarily or permanently) and to support people.
- Combining placemaking engagement activities with an activity/offer of social value and impact for local people.
- Supporting intergenerational dialogue around place, health and wellbeing.
- Building confidence in people as active citizens.

Granville New Homes

Between 2005 and 2007, The Glass-House worked with the London Borough of Brent to support a participatory design process for the Granville New Homes scheme in South Kilburn, which saw the development of 140 new homes of mixed tenure, the re-provision of the Tabot Youth Centre and the creation of a new pocket park (The Glass-House 2007). An early Glass-House project, this remains an excellent example of the value of design engagement and empowerment long after the builders leave.

The Glass-House programme for Brent Council provided a series of design workshops for a group consisting of members of the Granville New Homes Resident Steering Group (GNHRSG), the local New Deals for Communities team, and members of Brent Council staff engaged in the redevelopment. The

Figure 6.5 Granville New Homes Resident Steering Group members manning the public exhibition of design schemes – March/April 2005[2]

Figure 6.6 Residents and council staff on a study visit to the Netherlands

workshops focused on giving participants a shared understanding of the design journey ahead, and a shared language for working together to move the design process forward and included:

- elements of technical knowledge;
- ideas generation and collaboration;
- a study visit to the Netherlands to be inspired by different examples and approaches to housing design.

There was a clearly defined core working group that brought together council officers and local people to develop design quality indicators, set the design brief, interview potential design teams and construction companies, deliver consultation events and ultimately work with the design team, Levitt Bernstein Architects, throughout the iterative design journey.

The resulting Granville New Homes scheme has been widely acknowledged and awarded for its high quality design. The shared vision and commitment to providing homes that actively contributed to people's quality of life and to their health and wellbeing led to homes that are much loved and appreciated by the residents.

Several members of the Resident Steering Group remain in contact with the council and have continued to collaborate on other projects and to champion design quality and community engagement in placemaking in the Borough.

Since its completion, the Granville New Homes project has offered fertile ground to inform and inspire. The Glass-House, working with Brent Council, Brent Housing Partnership and Granville New Homes residents, has organised a series of study visits to help others learn from this exemplary project. Visitors have included housing and regeneration professionals from around the UK, community groups and organisations and even a group of planning and design officers on a visit from South Korea. The scheme stands out to visitors not only for its built outcomes but also as an example of a truly participatory and engaged approach to placemaking, which is evidenced by the ongoing commitment of those involved to sharing their experience and learning and to championing great design and this participatory approach to placemaking.

Feedback from a member of the Granville New Homes Resident Steering Group

If the foundation upon which a place is built is great, then the place will be great.

(Member of the steering group)

Key opportunities for social value:

- Empowering people to play an active role in shaping their built environment.
- Creating opportunities for capacity building and skills development through a participatory design process.
- Taking people outside of their own environment to be exposed to different places and approaches, and to be inspired.
- Improving the quality of places.
- Creating new champions and enablers of design and place quality.

Placemaking and media skills exchange

In partnership working with Silent Cities, a social enterprise in Sheffield, we created an opportunity to use media tools to explore the design of our places and to empower people to articulate views about their local built environment in new ways. This work formed part of a collaborative action research project called Scaling Up Co-Design (The Glass-House 2014c, 2014d) led by The Open University and funded by the Arts and Humanities Research Council. It brought together Silent Cities' Community Journalists programme that trained people facing silent issues from homelessness to mental health with digital skills with a Glass-House design engagement programme at a school in Elephant and Castle, an area in South London undergoing large-scale regeneration.

The first stage of this project was a Glass-House workshop for the Community Journalists in Sheffield exploring place quality and equality. Using their newly acquired media skills, participants used audio recording, photography and film to explore their local environment, articulate their relationship with it and explore how local places affected them as they moved through them. The media they

Figure 6.7 Photo taken by a student at Sacred Heart Catholic School during the media
       workshop

produced allowed them to explore and reflect on their environment and to voice
their perspectives in new ways. The resulting artistic pieces they created offered
a powerful representation of the city of Sheffield from different perspectives by
capturing the sensory impact of streets including personal commentaries on the
qualities and physical features of the city spaces that made them feel included or
excluded.

Two of the community journalists and the director of Silent Cities then
came to London to work with The Glass-House to deliver a media workshop
they had co-designed for students at the Sacred Heart Catholic School in
Elephant and Castle. This workshop was integrated into some work that The
Glass-House was doing for the development team from Lend Lease that was
leading a large-scale regeneration scheme in the area. Lend Lease were keen
to engage young people in a conversation about the new development and had
commissioned The Glass-House to help prepare a group of young people for
that dialogue.

The media workshop was an opportunity to bring together design engagement
in local development with training in media skills for the participating young
people at Sacred Heart Catholic School. Following an initial introduction to basic
design principles and to people's relationship with the built environment from
The Glass-House, the Silent Cities team quickly trained the students to use digital
sound recorders, cameras and video cameras. The students then moved around

A space for expression...

Figure 6.8 Still from a photo story produced by students at Sacred Heart Catholic School during the media workshop

their school capturing their impression of the various spaces within it, which they then brought back to the workshop to transform into artistic representations of their perceptions and ideas regarding the design of their school.

In the space of a day, the young people developed a new understanding of the design of the places around them, trained in media production and produced deeply insightful and moving works of art that they shared with other students, the development team and later, with their school community. The community journalists, who had only recently acquired those media skills themselves, moved quickly from the role of student to teacher, which was an empowering process for them, as well as hugely valuable for the young people involved.

The artistic outputs that both the community journalists and the students from Sacred Heart School produced demonstrate the empowering value of simple capacity building opportunities, as well as the power of collecting new and different views and voices about our built environment.

Feedback from participants

> We've learnt from them, they've learnt from us and the collaboration is quite a feat in itself to do between a Charity in London, a social enterprise in Sheffield, a mix of volunteers and a school.
>
> (Community journalist)

[We've learnt] how to use media to tell stories and how media can reveal the beauty in simple things we see every day.

(Student)

Key opportunities for social value:

- Building public awareness, interest and investment in built environment.
- Empowering people to articulate their views on their built environment.
- Creating opportunities for capacity building and skills development through design engagement process (for example creative arts practices, communication, life skills and employability).
- Empowering people to pass their newly acquired skills on to others.
- Taking people outside of their own environment to be exposed to different places and approaches, and to be inspired.

Invest in and learn from collaboration and experimentation

The four examples of placemaking projects included in this chapter have relied on people working collaboratively and often stepping outside of their comfort zones to experiment with new ideas and approaches. They have involved a willingness on the part of both project leaders and participants to challenge their own assumptions and those of others, to ask difficult questions and to explore and test new ways of thinking and doing. They also began with the recognition that experimentation brings risks. Indeed, in most cases, the projects showcased took some unexpected directions and hit bumps along the way.

However, if we are to achieve real social value and impact in placemaking, it is essential to work collaboratively and to find ways to mobilise and realise the potential of local people, places and assets. Empowering local people and organisations to contribute to any placemaking initiative requires the investment of time and resources and a willingness to experiment. This does bring with it some risk, but it can also pay high dividends.

While working to reach defined objectives, trying out new ways of working can lead to valuable discoveries and unforeseen positive outcomes. Indeed, any outcomes of collaboration and experimentation, whether perceived as positive or negative, can contribute valuable learning that if absorbed and implemented, can ultimately improve the end result. If shared, that learning can not only contribute to the success of an individual project, it can also help bring about valuable shifts in culture, policy and practice, at a community, organisational, regional or national level.

There is always a need to consider financial viability, cost-effectiveness and value for money in placemaking. Nevertheless, we should perhaps also agree that the parameters for investment and for measuring success need to be informed by the social value that any project, small or large, has to offer.

If we are to enhance the social value created through placemaking, we must be willing to collaborate, to experiment, to reflect, to learn and to share our learning with others. We must consider and assess the social value in placemaking at an individual, organisational and industry level. We must look at the impact of our built outcomes but also at the journeys that those involved in the placemaking process have madeand at the opportunities that the design and development process itself has offered.

We must constantly ask ourselves what we can do to explore and create new and better opportunities for creating social value through placemaking and how a more experimental, collaborative approach and willingness to share our learning with others can contribute to success. Learning from good practice, driving innovation, and sharing strong evidence and powerful stories within and across sectors has the power to change hearts and minds, and put simply, to help us do things better in the future.

Summary

In placemaking, the potential for social value lies not only in the built outcomes and their impact locally but also in the process of designing and transforming our places. Design processes that involve local people and organisations can unearth and mobilise knowledge, connections and assets and, at the same time, empower people with new confidence, skills, and social capital.

Empowering design practices can be applied to projects at any scale, whether led by community-based organisations, the public or the private sector. They benefit from clear strategies and from inclusive, transparent and accountable processes. Above all, they rely on a commitment to collaboration and a willingness to experiment.

We must constantly seek to enhance the potential for empowerment and social value creation in design and placemaking, measure the impact of projects on people and places, and capture and share our learning with others.

Notes

1 Photographs in Figures 6.1–6.4 and 6.6–6.8 are Courtesy of The Glass-House Community Led Design.
2 © South Kilburn New Deal for Communities.

References

The Glass-House (2007) *Granville New Homes Case Study*. London: The Glass-House Community Led Design. Available online at: www.theglasshouse.org.uk/project/granville-new-homes-south-kilburn.

The Glass-House (2014a) *White City Green Case Study*. London: The Glass-House Community Led Design. Available online at: www.theglasshouse.org.uk/project/white-city-green-empowering-young-people-through-new-development.

The Glass-House (2014b) *Tidworth Mums Case Study*. London: The Glass-House Community Led Design. Available online at: www.theglasshouse.org.uk/project/unearth-hidden-assets-tidworth-mums.

The Glass-House (2014c) *Unearth Hidden Assets, Through Co-design and Co-production Case Study*. London: The Glass-House Community Led Design. Available online at: www.theglasshouse.org.uk/project/unearth-hidden-assets-through-community-co-design-and-co-production-2.

The Glass-House (2014d) *Scaling Up Co-design Research and Practice: Building Community-academic Capacity and Extending reach*. London: The Glass-House Community Led Design. Available online at: www.theglasshouse.org.uk/project/scaling-up-impact-and-reach-through-co-design.

The Glass-House and the Open University (2014) *Tidworth Mums: A Case for Soft Play Resource Pack*. London: The Glass-House Community Led Design. Available online at: www.theglasshouse.org.uk/project/tidworth-mums-a-case-for-soft-play.

7   Meadowhall, Sheffield, UK

Co-creating social value on a large private sector project

*Ani Raiden, Martin Loosemore, Andrew King
and Chris Gorse*

This chapter provides an insight into how a large client and large contractor work together and with the supply chain to create social value on a successfully completed £60 million refurbishment project. It takes the form of a descriptive case study of the redevelopment of Meadowhall shopping centre in Sheffield, UK, which was completed in 2018. It makes extensive use of representative verbatim quotations from practitioners involved in the project, and interviewed for this case study, to report in their own words how parties coalesced to create social value. In addition to discussing this specific project, participants make reference to other less positive experiences related to social value on other projects that prove illuminating.

The project

Meadowhall is one of only six super-regional shopping centres in the UK, employing over 8,500 staff at peak times and attracting over 25 million visitors per year and is located on the site of a former steelworks four miles from Sheffield city centre. In the summer of 2014, almost 25 years after Meadowhall first opened, the decision to undertake a £60 million refurbishment and modernisation project was taken. The works, procured using a two-stage design and build route incorporating an amended JCT D&B contract, included the following activities:

- Diversion of Mechanical, Electrical and Public Health (MEP) services from walkways onto roof.
- Removal of service walkways and bridges.
- New high level wall cladding particular to each mall – replicated around voids in upper mall slab.
- General declutter of the malls – reduce column widths, remove 'fussy' detailing.
- Relocation of feature lifts to improve sight lines – Arcade and Park Lane Malls.
- New glass balustrading.
- New MEP services – high tech lighting, natural ventilation and smoke extract system.
- WC refurbishment.

The organisations

British Land

British Land, one of the UK's leading commercial property companies, jointly own Meadowhall in conjunction with Norges Bank Investment Management (the Norwegian Government Pension Fund). Valued at £1.9 billion, Meadowhall constitutes British Land's biggest retail asset in a portfolio that spans the UK and includes high quality retail, London offices, leisure and residential sectors. The organisation's size, scope and performance is notable with £13.5 billion of assets, £18.1 billion of assets under management and 25.2 million square feet of floor space utilised at a 97.6 percent occupancy rate (BL, 2017).

Laing O'Rourke

Laing O'Rourke is the UK's largest privately owned international engineering enterprise, working across a range of industry sectors. It was selected as preferred bidder at the end of the first stage of a two-stage design and build procurement route under a pre-contract services agreement. Following further design development and a detailed value engineering exercise, they entered into contract as principal contractor with BL in December 2015. Innovators in digital engineering and off-site production through their approach to Design for Manufacturing and Assembly (DfMA), Laing O'Rourke are unusual in having developed a mature vertically integrated supply chain, where a number of key services are provided 'in-house'. Their building and infrastructure services specialists, Crown House Technologies, carried out a significant part of the works at Meadowhall.

In addition to their vertically integrated supply chain, Laing O'Rourke relied on a number of external companies in their supply chain to help develop and deliver the project. For the purposes of this case study, focus is given to two specific local SME's: Clearline and Evergrip.

Clearline

Clearline is a Sheffield-based contractor specialising in curtain walling, structural glazing and building envelopes. They provide an installation and refurbishment service across the UK and Europe from their Sheffield and London sites. On the Meadowhall project they were employed as a specialist subcontractor by Laing O'Rourke to carry out the reglazing works. This package, worth in the region of £5 million, included replacing the existing rooflight glazing that required stripping back to structural steelwork prior to replacing with new units. Clearline employed three new apprentices specifically for Meadowhall.

Evergrip

Evergrip specialise in the design, manufacture, import, fabrication and installation of glass reinforced plastic (GRP) composite safe access products across the UK. On Meadowhall, they worked with Laing O'Rourke to relocate the services to the roof. Prior to refurbishment, the Meadowhall centre utilised service walkways in the centre that were used to route the services and allow maintenance staff access. As part of the redesign, these were removed and replaced on the roof using a value engineered GRP 'services raceway' made from a lightweight bespoke solution that was delivered to site in kit form for assembly. The new bespoke design, manufactured in China, drastically removed the need for around 450 surface penetrations with potential for water ingress, reduced the weight of the structure, allowed swift installation in addition to reducing trade interfaces.

The case study participants

This case study draws on interviews with a number of individuals involved in the Meadowhall project.

Lesley Giddins, Director, Sandgrown Consultancy
Lesley is an experienced practitioner who has worked in both the public and private sector and supports British Land's wider sustainability strategy on Meadowhall. Her experience is informed by her current role working on a range of projects across the country for British Land and other clients, and she acted as a consultant for them on Meadowhall.

Dave Higgins, Senior Project Manager, Clearline Ltd
Providing daily operational management of Clearline's package of works on Meadowhall, Dave was responsible for interacting with Laing O'Rourke on a daily basis during the works, in addition to overall management of Clearline's apprentices.

Matt McKirgan, Community & Regeneration Advisor, Laing O'Rourke
Matt has led Laing O'Rourke's sustainability strategy for Meadowhall, requiring him to interact with a wide and diverse range of stakeholders. He began working on the project after works had already commenced on site.

Gayle Morgan, Senior Commercial Manager, Laing O'Rourke
Gayle led financial and commercial management operations for Laing O'Rourke on the Meadowhall project. She has been involved in the project throughout Laing O'Rourke's involvement, and her work in this case study focuses on the tender and procurement process at main contract and subcontract level.

Nick Osborne, Managing Director, Evergrip
Nick established Evergrip in 2001. He made initial contact with Laing O'Rourke as a new client on the Meadowhall project, secured supply chain status with the organisation, managed the procurement and value engineering and maintained oversight of the project in conjunction with his operation team.

Nick Slater, Senior Facilities Manager, Sheffield City Council
Nick works in the Transport and Facilities Management department at Sheffield City Council, and his work as Senior Facilities Manager on this project involved him providing council overview on the redevelopment of Wincobank Community Centre.

James Stringer, Glazing Installation Apprentice, Clearline Ltd.
James worked on the installation of the new glazing units on Meadowhall. Following school, he returned to college to retake GCSE maths and English. He then studied plumbing at college before leaving mid-course to earn money working for a data cabling company beforestarting work and his apprenticeship at Clearline.

The following themes draw from the case study research carried out over a two-year timeframe from the project commencing on site in January 2016 to its completion two years later. Eleven semi-structured interviews were carried out during this time in addition to a number of site visits.

Key themes

Values and priorities

For British Land, sustainability sits at the core of their values. Their 2020 Sustainability Strategy 'We create Places People Prefer' recognises their role in developing and maintaining a significant architectural footprint across the UK. The strategy, developed through focused stakeholder engagement including a range of different groups including local communities, supply chain, investors, internal staff and local authorities, is underpinned by four pillars:

- Wellbeing;
- Community;
- Skills and opportunity;
- Future proofing.

These values help shape the way that social value was created at Meadowhall. As many of their assets, including Meadowhall, are held over the long-term, this allows them to engage with a range of stakeholders on an ongoing and project-specific basis. Shoppers, retailers, local and national organisations, local community groups, the local authority and their supply chain working group all fed into the consultation for the Meadowhall refurbishment. Lesley Giddens highlights the benefits of this collaborative approach:

> What that creates is buildings that are then used by people because they've had agency in saying what it is that they would like to see on the site. British Land do annual surveys on the site about what people want and what they would find helpful.

For British Land and Laing O'Rourke, this approach to consultation played a large role in the project being deemed such a success on a range of performance criteria. Laing O'Rourke work for a range of predominantly large clients across the world, the need to take account of their clients needs and the importance of being perceived as a good employer are key, as Matt McKirgan points out:

> It's of growing importance, particularly for a company like Laing O'Rourke who now almost exclusively deal with very large projects, high value and high profile clients, they have a very keen focus on sustainability, its about their image, their place in the world which they are trying to affect, so we need to fall in line with that, to show that we are sustainable and fall in line with our clients, but again Laing O'Rourke is good at seeing beyond the books, its about being a positive employer.

On the Meadowhall project, this involved them helping develop apprenticeships in their supply chain and providing work experience opportunities. In addition, they educated school children and university students about the project and opportunities in the construction industry, supported British Land's charity of choice, Bluebell Wood, and refurbished Wincobank community centre with a focus on whole-life costs.

As a large national organisation, British Land have developed a range of projects with local and national agencies including the National Literacy Trust, and also run autism and dementia projects at many of their major assets. This balance of local and national is considered key in maximising the social value they can deliver across their significant portfolio.

Laing O'Rourke similarly have a well-developed and embedded approach to social value that has recently been updated in their new social sustainability framework (SSF). Inside the business, social value is embedded as part of the human capital function, which is seen as giving it the ability to influence the wider business. Interestingly, Laing O'Rourke have found that young people in the business are particularly socially aware, with this being attributed to how they are educated at school and particularly at university.

The participants' high social value orientation was a thread that ran throughout the Meadowhall project with those involved in the case study all taking personal satisfaction from their work as Lesley Giddens points out:

> I come from a background that tells me you should treat people well, you should treat your employees well and you should train them well. A well-trained workforce is a more positive workforce.

For Matt McKirgan from Laing O'Rourke, the desire to have a positive social impact lies at the heart of his work:

> As an employee, I want my firm to be sustainable, part of that is being well regarded, that we are ethical. It fits in with my philosophy of life in general;

no matter what sector I was in I have that view. There's no medals or things like that, but you can feel you are having a positive impact.

The intrinsic personal satisfaction that is often derived from social values activities often stems from first-hand involvement. Lesley Giddens sees the need to continually encourage people to participate in the creation of social value as key:

> I feel my own personal job in this regard is to just keep nudging them and try and further open their eyes to the value of this for their own company, for their own wellbeing. I've seen hard-bitten construction guys working in schools and initially they don't want to do it, but actually they love it when they get in there with kids who are excited by it all, by having them enraptured by trying to build a brick wall!

Employment and training

Employment and training related issues impact in a number of ways: the importance of apprenticeships, skills shortages, young and old workers and the need to train on an organisational level. Of overriding importance is the agency that all participants felt in driving the development of appropriate skills in their businesses and supply chains. Set against a back drop of a UK construction skills shortage and an increasingly ageing workforce, fears over the impact of Brexit are leading to an increased need to attract new entrants to the industry as Lesley Giddens clarifies:

> I think construction is starting to change its own views because the skill shortages are now biting. They are no longer theoretical shortages that are coming in the future; they are manifesting themselves in a difficulty in finding companies in the supply chain and suitably qualified individuals.

For Laing O'Rourke, the skills shortage is a similar driver:

> Laing O'Rourke have always tried to be an innovator in the sector, at least be at the forefront of developments, there is that selfish element, we recognise there is a skills shortage, which needs addressing, if we want to be a leading engineering company, we need to make sure we have the right skills and help facilitate that, nobody will do that for us, so we need to take the bull by the horns to develop our people and the people to take their place in years to come.

All participants believed that it was firmly in their interests to develop the workforce and attract new entrants. For Dave Higgins from Clearline, developing their own skills in-house through on the job training is key in an industry where many organisations take a short-term profit driven approach with a corresponding impact on quality:

In our industry, curtain walling, there isn't a great deal of formal training about, so the only way we can make sure that we've got the right kind of guys is by training them in-house, so we are more active in that respect... because of the lack of training in our industry, getting an apprentice trained to the right level is key, there are so many corners that can be cut in our industry, which people can get away with doing, you need to try and train that out of people and the very best way to do that is right at the beginning isn't it?

Apprentices were universally seen as key to developing employment and skills, with British Land aiming for 3 percent of their supply chain workforce to be apprentices by 2020. It is important to note that all 24 apprentices employed on Meadowhall were craft based and located in the subcontractor supply chain. As Laing O'Rourke are a national employer, the practicalities of moving apprentices to different projects as the work allows creates difficulties that are overcome by utilising their supply chain:

Apprentices, by nature of the age range and pay, are not as likely to want to travel to another Laing O'Rourke job 200 miles to finish their apprenticeship. We are luckyish here that it's a 2-year job so you could theoretically get through an apprenticeship, but that's why its easier for the supply chain to recruit the apprentices so there is more stability there, it's the geographical importance of the apprenticeship, you could say they have the added value of working on a number of different projects, so that broader experience. LOR does have lots of apprentices, but they are on longer-term and larger projects and there is a whole host of training and support in place for them.

(Matt McKirgan, Laing O'Rourke)

The overriding industry focus on supporting young people has started to shift to supporting older people to retrain and return to work demonstrating the shifting sands of social value impact. For Lesley Giddens, British Land's long-term collaborative approach pays dividends in responding to the market:

Older people coming back onto the labour market for example are seen as more of a priority by a growing number of local authorities ...because there has been a bit of a drop in unemployment, the pool of young people who are unemployed and seeking work and anywhere near the labour market is getting smaller, so British Land are constantly about thinking about what's the best way they can deliver social value?

The drop in unemployment amongst younger age groups sits alongside a lack of young people entering the industry. Whilst much discussion at national level revolves around improving the 'face' of construction to make it more attractive to young entrants, Lesley Gidden's work interacting with a range of schools tells a different story; one of a lack of awareness of the variety of roles available. As such, education and awareness raising is key:

> They frequently don't so much have negative views as much as not really know what it means, you know, it means brickies, they don't think about designers, the office jobs, all the other people involved in it. So working with schools, working with young people to encourage them to try different things, working with older people coming back onto the labour market to get them to think about construction and actually suddenly your social value becomes not just a lovely cuddly thing to do, but actually an imperative and that's when the cynics start to think 'you know what, actually I can see the value in this'.

The belief that many potential entrants to the construction industry are simply not aware of the opportunities available is not solely limited to non craft-based roles. For example, James Stringer who is serving his apprenticeship with Clearline had the following to say:

> I do rope access training now and I'd never heard of that in my life before I started at Clearline, and I did see it and what they were doing up there, and I took a fancy to it really.

For James, who left school without any formal qualifications (GCSEs) before returning to college to take maths and English, his apprenticeship is paying real dividends:

> It's absolutely brilliant, from what I knew when I started that was about one and a half years ago its absolutely brilliant really, I feel like I am really getting somewhere… I am starting to get there, and I want to do my level two very soon in rope access and I'm going onto a new job soon to do cladding, which is obviously different to glazing, but it's the same type of work.

As can be seen, the ability to demonstrate tangible benefit for the wider business and industry is often seen an important way to educate people about the benefit of social value oriented activities.

Section 106 agreements

One area that is deemed particularly important in terms of social value is given is the local authority planning process; specifically Section 106 (S106) of the Town and Country Planning Act (1990). In the case of the Meadowhall refurbishment there was no S106 requirement, but in line with their own policies, British Land developed a Strategy that outlined commitments to local employment and supply chains. Various themes related to S106 agreements emerged as part of this work and are given significant space in this chapter.

Sometimes referred to as 'developer contributions', S106 agreements are often a key milestone in the commissioning cycle. These private agreements between developer and local authority are used to make a development application acceptable and are used when they are:

- needed to make the development acceptable from a planning perspective;
- fundamentally related to the development;
- 'fairly and reasonably related in scale and kind' to the development.

British Land take a very proactive approach to the S106 agreement and in this regard aim to engage early with local authorities by asking 'what would you like to see?' Their aim, as with all projects, is to develop a meaningful conversation and a 'shared problem' that can translate into tangible and significant social value impact.

The majority of the work was carried out at night on the project to avoid impacting on retailers and shoppers. This led to an understanding that it would not be possible to employ apprentices on site for much of the work as this was undertaken overnight as clarified by Lesley Giddens:

> It was agreed that the majority of the contractors could not employ apprentices as the work was done overnight. What is the issue here? Young people cannot be employed on night work in this field), so if they were to just enforce the number they would not be achieved. In all relevant cases we try to have a reasonable discussion with the Local Authority In fact we try and do that more and more and sit down and say 'can we look at what this development actually is and what are the meaningful things that we could do?' ...What we try and do is create a shared problem, it's not people on the other sides of the table measuring the success of the other, what I feel is that we should all care about young people getting jobs here so we want to consider what is the best thing we can do to make that happen.

This type of discussion is key to understanding the local authority drivers and is followed by a detailed desktop study to investigate the authorities' policies, procedures and plans to gain a more detailed insight into their needs and aspirations. In addition to the desktop study, local and national agencies are consulted. In this instance the employment and training plan that British Land developed was then reviewed internally to explore its fit with their strategy and procurement process. The Meadowhall plan includes:

- Employment and training requirements – with targets for a percentage of local people in the workforce. as the work on Meadowhall was mostly carried out at night, this impacted on apprenticeship requirements.
- Procurement of supply chains – with local targets.
- An engagement with local communities and schools as appropriate.
- Community development – the refurbishment of the lower floor of Wincobank village hall community centre.

The Wincobank project was selected as the key focus for community development through the centre's long-standing relationship with the local community. Nick Slater from Sheffield City Council worked closely with members of the Wincobank Village Hall Trust, Darren Pearce who is the Meadowhall Centre Director, Laing O'Rourke and the Sheffield City Council's designers. Sheffield City Council had already refurbished the upper floor of the centre, and against a backdrop of funding cuts and limited resources, Laing O'Rourke started early discussions with Wincobank community representatives as Nick Slater points out:

> They visited the building a number of times and talked to the community about what they wanted from that lower floor refurbishment. The council have had to say 'we have no further resource to do anything'. Laing O'Rourke, as they have gone along, have increased their resource into this local Wincobank community project and they have completed the specification, they have gone to building Regs with it and had discussions with planning where necessary ... Laing O'Rourke brought their designers in as well as their structural engineers to talk about the possible changes and what they need in their own resource package.

The collaborative and proactive approach that British Land take means that the local authority is often pleasantly surprised by their submission, and the work done on the refurbishment through the strategy has provided significant evidence to the local authority of British Land's credibility in this area. Another aspect that helps their success in this area is their ability to evidence a proven track record of creating social value impact on their previous projects.

It is worth noting that in Lesley's wider experience, and looking outside of the Meadowhall project, some local authorities can be overly focused around the application of their own targets and aspirations rather than looking to develop meaningful criteria; something that is believed to be impacted by their challenging budgets and associated resources. If rules are applied indiscriminately they can lead to less successful projects. In contrast, meaningful conversations and co-creation of criteria can lead to the authorities – and more importantly local communities needs being met in different ways. Nevertheless, there can sometimes be a reluctance to enter into dialogue by authorities. For example, helping local SME's grow and develop is key to British Land:

> In procurement terms British Land have moved much more to look at growing SMEs and thinking about how they can support people and this is a central tenet of what they are trying to do in terms of employment and supply chains. Considering 'how can we support those people who wouldn't naturally get the job or get the employment?'
>
> (Lesley Giddins, Sandgrown)

In addition, as local authority catchment boundaries can intersect individual streets, situations can arise where training courses can be made available to people in some dwellings in a particular street and not their neighbours in the same street.

This is a significant issue and whilst these situations can be overcome, they require local authorities to avoid blindly applying rules and regulations.

British Land have encountered other instances with other local authorities where they have not requested any specific requirements for local employment. In these instances, British Land will still include their own project specific targets based on their company values and stakeholder consultation.

Tendering, monitoring and measuring

The two-stage tender allowed a more collaborative approach to scheme development that impacted on the creation of social value. Interaction between British Land and Laing O'Rourke starts in the first stage when British Land provided the tenderers with a simple one-page document outlining the key sustainability drivers, which represent the key aspects of the British Land requirements (or in other cases of S106 requirements). These drivers are used as a starting point for further discussion and clarification and following Laing O'Rourke's appointment as preferred bidder during the second stage are supplemented with a comprehensive 40-page project-specific sustainability strategy outlining the local procurement, employment and training and community project drivers. This document forms the basis of Laing O'Rourke's more detailed response as Gayle Morgan from Laing O'Rourke describes:

> One of our bid team at the time sat with their equivalent and went through that document line by line and some things changed, I think some things reduced and some things went the other way where we felt the targets were a bit low and could definitely be better we could say that was modified. The final agreed version is part of our contract and we are monitored against each of them.

As can be seen, mirroring how British Land work with local authorities, a collaborative approach is taken to developing the formation of the contract, with detailed and meaningful conversations underpinning the work. Just as British Land often go beyond the requirements of the S106, Laing O'Rourke often go beyond the clients social value needs. In this way, they are responsive to the clients needs and also embrace their organisational social values and sector specific requirements.

> Its got to be a collaborative thing, generally speaking … you've got to be selective on what you do and collaborate with a client. For a client to say arbitrarily these are your targets I don't think works as there are so many things that could impact on your ability to meet those targets that you've got to give it some careful consideration … We have very similar [key performance indicators] KPIs, but our targets are actually higher in most areas, not all, take apprenticeships for example, I can't remember what British Land asked for, but our own internal targets were higher than British Land for apprenticeships.
>
> (Gayle Morgan)

Laing O'Rourke similarly pass their social value requirements through their supply chain. Generally commencing in the early part of the second stage, they articulate their needs, for example apprentice requirements, in their supply chain tender documentation. Four of the 24 apprentices on Meadowhall were classed as 'new', with the rest classed as 'sustained', as the apprentices were already employed in the supply chain. Owing to the changing definitions of apprenticeships introduced by the government, a much wider range of in-house training is now included within the definition, meaning targets can be reached more easily. British Land are acutely aware of this issue and whilst they can see the benefit that the wider definition can have in encouraging various types of training, they are keen to protect the value of the term and as such have developed a much narrower definition to underpin their own metrics:

> We've separated them out in our own reporting to say that bringing new people into the industry is one thing and training older people, or indeed young people who are already in the industry is a separate, but equally good thing.
>
> (Lesley Giddins, Sandgrown)

The project does not make use of any type of social return on investment (SROI) model. Instead, social value is measured on a range of criteria throughout the project. British Land both report to the local authority on the agreed criteria and have an internal reporting system that allows key metrics to be shared amongst different functions and inform decision-making. Similarly, Laing O'Rourke report on a monthly basis to British Land and also internally on their impact reporting system.

British Land also carry out additional routine audits of Laing O'Rourke through office visits and documentary analysis. Embedding social value in the procurement system, developing meaningful metrics and developing a simple reporting procedure into the project management system is key to the smooth running of the project. It is not unknown for social value to be considered an afterthought on some projects, meaning site teams have to exert pressure to meet targets. For Matt McKirgan, the way in which Meadowhall's system was developed early in the project paid dividends:

> I've been really lucky as things have been established so well at the outset, our outcomes have just been developed naturally so its been brilliant, probably a model to advocate as its so easy on resource, I haven't had to spend lots of time driving this to happen.

Flexibility in the ability to create social value is seen as key for British Land and Laing O'Rourke. As described, engaging local authorities in meaningful discussion of a shared problem can allow creative approaches to be developed.

> the more people at different levels of the chain involved the better. It could be as simple as a major order for a local firm, an SME who may need to upscale their workforce, do you leave them to do that or do you help them upskill their existing staff and new staff opportunities?
>
> (Matt McKirgan, Laing O'Rourke)

Both British Land and Laing O'Rourke are increasingly focused on how a positive legacy can be created through working with smaller local organisations to help develop their businesses. One example from Meadowhall is Laing O'Rourke working with Evergrip on the relocation of the services to the roof of the centre. Constituting Evergrip's largest order to date, their managing director, outlines how working with other large contractors can often feel:

> Sometimes other contractors work in such a way that it feels like they are doing you a favour to work with them...being approached by a large organisation can be overwhelming and being over-keen can lead to mistakes. What we have found is that LOR will stop and listen, they take notice and cooperate and work with us.
>
> (Nick Osborne)

In contrast, they felt that Laing O'Rourke took them seriously from the start and gave them support; an approach that gave Evergrip the confidence to develop a successful approach to working with such a large organisation. Moreover, they found that Laing O'Rourke acted differently to how they have experienced some other large contractors during the value engineering exercise as Nick Osborne points out:

> When we were initially approached by Laing O'Rourke we ran our engineering calculations on the product we developed and it was taken very seriously, we had a number of meetings here at the factory and there was a definite appetite to solve a problem ... work we did in the background to solve the problem could have easily been taken by them to one of our competitors to see if they could beat us, but there was none of that business at all.

Sharing ideas at tender stage, both at client-main contractor and contractor-subcontractor can help to stifle innovation; a problem that is all too familiar in an industry sadly still often characterised by a short-term approach. For Nick, there were other advantages too:

> They were absolutely first rate in terms of speed of payment, information sharing, cooperation, they ere very professional to deal with and for us as a small local company it gives us confidence to show ourselves and our other customers what we can do.

As with Evergrip, the Laing O'Rourke subcontract order was the largest to date for Clearline. They felt supported in a different way; to help develop apprentices through on the job training as Dave Higgins clarifies:

> I can only give you my judgement based on this Laing O'Rourke site team, they have been brilliant, one of the best site teams I have worked for, they are very proactive When it comes to education. You see a lot of the site team

are absolutely programme-led and they probably wouldn't like the idea of us training guys up on site as it exposes them to potential project delays; if we have inexperienced guys and we have to slow down to teach them, a lot of site teams don't actually like that, but Laing O'Rourke embrace that and they were actively buying into that really.

As can be seen, the challenges of on-the-job training are made much easier to navigate with an encouraging client. The benefits of this type of experiential learning are similarly clear as James points out:

> At the beginning they showed me the basics of it and what I needed to do to make a start and then they left me to it, they let me do the jobs so I could learn it myself, because you are better off doing it yourself than watching someone if you ask me because you are going to learn it a lot quicker and that's' how its gone. If I need to ask a question I'll ask, you're better asking a question than making a mistake because it could cost a lot of money or take time.

Summary

This case study demonstrates how a large, experienced construction client, large contracting group, local authority and the supply chain coalesce on a large project to develop social value with tangible impact. The importance of carefully developing appropriate values and priorities was underlined as key in delivering employment and training benefits. The client's long-term involvement in Meadowhall allied to its deep-rooted place in the local area paid real dividends in understanding what important to local people. Focus was similarly given to the section 106 agreements and how this can be successfully developed through a careful and collaborative approach formed around meaningful criteria. Finally, tendering, monitoring and measuring were explored demonstrating how social value is developed on a two-stage design and build tender including local procurement of a high performing supply chain leading to real training and education opportunities for local people.

8   Multiplex's Connectivity Centres©

An exemplar of social value in action in Australia

Dave Higgon and Joanne Osborne

This case study describes the development, operation and social impact of Multiplex's Connectivity Centres© in Australia. The Connectivity Centre© concept is a collaborative initiative, which has been developed, tested and refined over many years by Multiplex in collaboration with its clients, its supply chain and community, voluntary, not-for-profits and third sector organisations working on its projects. Multiplex's Connectivity Centres© are an innovative response to the social value requirements of its clients and the communities in which it works and is a central part of its wider Linking Industry Needs to the Community (LINC) programme which aims to leave a positive legacy in the communities in which Multiplex builds.

What is a Connectivity Centre?

The overall aim of Multiplex's Connectivity Centres© *is* to leverage the opportunities presented by the construction and operational phases of its projects to support local disadvantaged and unemployed people through the provision of structured training and employment opportunities. Connectivity Centres© not only provide a pathway to employment for these local people through workplace training, job placement, employment and wrap-around support services, but also provide whole-of-community resources such as space for local community groups to support Multiplex projects to deliver community benefit.

Multiplex Connectivity Centre© harness the power of the construction industry and the process of cross-sector collaboration and social procurement to build innovative, productive and resilient communities while also building a diverse construction workforce drawn from all members of society irrespective of background. Multiplex builds in some of Australia's most disadvantaged communities and it recognises the huge potential role it can play in providing a pathway to work for society's most vulnerable groups.

While the creation of sustainable employment is central in resolving many community challenges, Multiplex Connectivity Centres© are not just about employment, but more broadly about providing 'social value' by addressing a variety of community challenges through a range of programmes. Multiplex defines social value as all types of value that its businesses, projects or programmes create, beyond just economic impacts to include social, environmental, cultural and health impacts at both a community and individual level.

Multiplex's Connectivity Centres©, developed and refined over a period of ten years, are an innovative example of cross-sector collaboration in action because they provide a physical and virtual space for organisations from the government, business, third sector, community and traditional not-for-profit and charity sectors, to collaborate and co-create innovative, cross-sector solutions to pressing social challenges in the communities in which Multiplex builds. Different cross-sector configurations, relationships, collaborations, joint ventures and partnerships, developed and tested over many years, are forged in different project locations in response to local community needs and priorities. Importantly, the community needs represent the starting point for formation of each individual Connectivity Centre© and the ultimate standards by which their success is measured.

Context

Multiplex's Connectivity Centres© have been developed within the context of important changes in Australia's social policy landscape and are designed to have an impact beyond the immediate vicinity of its project's communities to help address deeper structural challenges in Australia's welfare system. Contemporary developments which Multiplex were cognisant of when designing their Connectivity Centres included:

- Intransigent and growing social disadvantage and inequity in many parts of Australian society which seem resistant to traditional social welfare interventions.
- Trends in 'new public governance' which involve paradigm shifts towards relational and outcomes-based procurement involving the co-creation of public/social value through cross sector partnerships between government, private, third and community sectors.
- Increasingly wicked social/environmental problems, which governments cannot solve alone and require new innovative and entrepreneurial thinking.
- Growth of public/private sector 'social procurement' policies (Federal Indigenous Procurement Policy (2015), NSW Aboriginal Participation in Construction Policy (2015), QLD Building and Construction Training Policy (2015), QLD Charter for Local Content, QLD Buy Queensland procurement policy (2017), Victorian Major Project Skills Guarantee etc.
- Changing notions of 'value for money' among its clients (public and private), which increasingly incorporate concepts of 'social value'.
- Growing corporate social responsibility expectations of business by clients, communities, shareholders and employees.
- A growing third sector of socially responsible businesses to procure from (supply chain diversification).
- Innovation points in New Generation Green Star Rating Tools that recognise the importance of social value creation.

- Dysfunctional and highly fragmented job services sector which is incentivised to compete rather than collaborate and has few historical connections with, and understanding of, the construction industry.
- A highly fragmented, transitionary and commercial construction industry, which has historically seen the community as a risk rather than an asset and which has few connections and little understanding of the job services sector.

Programme philosophy

The philosophy underpinning Multiplex's Connectivity Centres is based on four key principles:

Community-driven

The community and its social needs, priorities, strengths and weaknesses represent the foundation and starting-point of every Connectivity Centre. These needs not only define the objectives and intended social impacts of a Connectivity Centre but the standards by which success is ultimately measured.

Shared value

The concept of shared value is based on the premise that sustainable solutions to social problems lie at the intersection between business and community interests and that there is a mutual interdependency between business and community interests. All organisations generate social, environmental and economic value and that the concept of value is inherently whole and should not be considered in separate parts, but as a single 'blended value' proposition. This position contrasts with the triple bottom line thinking adopted by many companies which inherently assumes that organisations have to make trade-offs between economic, social and environmental goals.

Collective impact

The 'collective impact' approach is based on a system's approach which asserts that that many of today's social challenges are too complex for any one organisation to solve alone. Collaboration across government, business, not-for-profit and community sectors is therefore critical to achieve significant and lasting social change. Unlike traditional social partnerships and alliances, collective impact initiatives are based on the formation of stable, long-term and high trust relationships across government, industry, third and community sector organisations underpinned by five main principles:

1. A common agenda for change among all key stakeholders including a shared understanding of the problem and a joint approach to solving it through agreed upon actions.

2. A consistent process for collecting data and measuring results consistently across all the participants which ensures shared accountability.
3. A clear plan of action that outlines and coordinates mutually reinforcing activities for each participant.
4. Open and continuous communication between all key stakeholders underpinned by trust, mutual objectives and a common motivation.
5. A backbone organisation with dedicated staff to coordinate participating organisations and agencies.

Wrap-around services

A Connectivity Centre© provides 'wrap-around' services to help the project community's most disadvantaged people transition to work. These wrap around services are focused on client and other stakeholder needs before, during and after work-placement and are designed to ensure vulnerable clients and Multiplex's business partners receive the full support they need at all stages of the transition-to-work journey. By providing a one-stop-shop where all of these services are coordinated, Multiplex's Connectivity Centres© overcome the problems of fragmentation which plague the job services and construction sectors and prevent collaboration and sustainable employment solutions being developed and implemented effectively.

Benefits

At the highest and broadest level, and in alignment with Multiplex's shared-value philosophy, the main benefits of Multiplex' s Connectivity Centres are as follows:

- For the community in which Multiplex builds – improved community resilience and leaving a lasting positive project legacy through the provision of 'sustainable' employment/training opportunities for disadvantaged people and vulnerable groups; and support for local businesses, third-sector organisations and community organisations and support networks.
- For individual Connectivity Centre clients – all benefits associated with sustained employment including: increased hope, self-respect and confidence for the future; increased integration, contribution and belonging to normal society; improved mental and physical health; improved wealth; reduced dependence on social security and welfare support services; reduced crime and re-offending; reduced substance abuse, suicide and family violence; reduced antisocial behaviour; positive spill-over multiplier benefits into families and wider communities.
- For government – innovation in achievement of social policy objectives; better value for money in procurement spend; increased tax income, productivity and innovation through higher employment and entrepreneurship and greater utilisation of disengaged and disadvantaged youth and workers; reduced costs associated with welfare, crime and health.

- For Multiplex, its business partners and clients – demonstrable innovation in Corporate Social Responsibility; improved employee recruitment, engagement and retention; compliance with growing social procurement requirements; improved community engagement and public relations; greater innovation in bids; positive reputation; and competitive advantage.

Measuring the social impact of Multiplex's Connectivity Centres©

Multiplex's approach to social impact assessment is based on the following best practice principles:

- simplicity – relevant, simple and easy to understand;
- value – helpful to enable real decision making and learning around the direction of resources to maximise social impact;
- proportionate – proportionate to the data available, nature and scope of the intervention, the time and resources available, capabilities of the organisations using the results;
- rigorous – based on a rigorous multi-method approach which reflects all stakeholder perspectives and is based on a range of tools and data, both quantitative and qualitative;
- bottom-up – driven by community needs, based on the perspectives of key stakeholders and accepted by them as a true representation of social impact.

Multiplex also adheres to the seven principles of good impact assessment, which are widely accepted as best practice in the international literature. The following sections describe how they are applied in a Connectivity Centre© context:

1. Involve stakeholders – stakeholders are involved in all assessments because they are best placed to describe the changes that have happened in their lives as a result of each Connectivity Centre© programme. In line with the philosophy of 'shared value', assessments of social impact also include other business and not-for-profit project stakeholders, while recognising that ultimately the focus is very much on the participants of the programme and their families and communities.
2. Understand what changes – the social impacts evaluated are evidence-based and supported by empirical research data. The changes evaluated include immediate, intermediate and long-term and both positive and negative. Impacts and reported for a number of key stakeholder groups.
3. Value the things that matter – most social impact reports are based on an external assessor's perception of value (often a community outsider) rather than on community stakeholders' perceptions of value. However, the imposition of social impact by external assessors perpetuates long-standing problems and feelings of oppression within disadvantaged communities by disempowering them and reinforcing differences that promote deep-seated inequity and disadvantage. For this reason, a triangulated approach is

employed using both self-assessment and peer-assessment in its social impact measurement framework.

4. Only include what is material – Multiplex will only report social impacts that can be 'evidenced' by primary and secondary data and collected using an ethnographic approach which involves an innovative mix of qualitative and quantitative methods (interviews, diaries, participant and non-participant observation, surveys, documents, focus groups, film etc). Data collected from a range of stakeholders before, during and after each Connectivity Centre programme is then analysed using thematic analysis and descriptive statistics and triangulated and cross-referenced to produce an unbiased, fair, balanced and accurate account of impact produced from the perspectives of those involved in and affected by our Connectivity Centres.

5. Do not over-claim – Multiplex will only report social impacts that can be attributed to its Connectivity Centre activities and will take into account various counterfactuals which include: deadweight (what would have happened anyway); drop-off (reducing benefit over time); attribution (what else could have contributed to the change); displacement (what other benefits does the intervention displace/push aside); and substitution (replacement of other gains).

6. Be transparent – Multiplex works hard to ensure that it is being completely transparent its approach requires reporting both positive and negative changes, revealing all data collection and analysis methods, data sources and limitations.

7. Verify the results – Verification of the results presented in a social impact report involves Multiplex providing an opportunity for stakeholders to comment on the results before they are included. This ensures that each report accurately reflects their view of reality regarding the changes in their lives, which have occurred as a result of being involved in the Connectivity Centre.

Social impact assessments involve collecting data from people and potentially vulnerable groups and raises important ethical questions. For this reason, all social impact assessments abide by the following basic ethical principles of research:

- honesty (does not fabricate, falsify, or misrepresent data);
- objectivity (avoids bias and discloses personal or financial interests);
- integrity (keep promises and agreements; act with sincerity);
- carefulness (designs assessment process to avoid error);
- openness (shares data, results, ideas, tools, resources);
- respect for intellectual proper (gives proper acknowledgement or credit for all contributions to the assessment);
- confidentiality (protect the anonymity of respondents and confidential communications and data);
- respect for people (respect your respondents and colleagues and treat them fairly);

- social responsibility (prevents harm to the community and to respondents and colleagues);
- non-discrimination (avoids discrimination on the basis of sex, race, ethnicity etc.);
- competence (informed by good research practice, professional competence and expertise);
- legality (complies with relevant laws and institutional and governmental policies);
- care (shows proper respect and care for people and animals);
- protection (minimise harms and risks and maximize benefits to human respondents/subjects and respects human dignity, privacy, and autonomy; taking special precautions with vulnerable populations; and striving to distribute the benefits and burdens of research fairly).

Multiplex builds in many Indigenous communities and is especially aware of the need to ensure that its social impact reports are conducted in a way which is sensitive to Indigenous cultural values. To this end, Multiplex also adopts the following strategies when conducting its social impact assessments in Indigenous communities:

- considering the needs, customs, and standards of indigenous people and their communities;
- avoiding poor research practices which have been shown by researchers to have been harmful in the past;
- consultation with and involve indigenous people and Elders in the research process;
- incorporating indigenous values and beliefs into the research design;
- developing a relationship with the indigenous community;
- researchers identify themselves and their background;
- giving research participants the opportunity to use their own preferred methods to voice their experiences;
- collective ownership of the research process and outcomes;
- agreeing on the dissemination of results;
- evaluation through community feedback;
- respecting community's past and present experience of research;
- recognising the diversity of indigenous Australian populations;
- recognising social, historical, and political contexts which shape experiences, lives, positions and futures;
- privileging the voices, experiences and lives of Indigenous people.

Multiplex continues to develop a rigorous, auditable and consistent approach to understanding, measuring and documenting the social impact of its Connectivity Centres©. This is informed by its own research in collaboration with leading universities, internationally published research and a variety of leading and respected social impact guidelines which are recognised internationally as best practice. The measurement process is undertaken independently by internationally leading university researchers which ensures that the reports of social impact produced are unbiased, verifiable and auditable.

Social impacts

In understanding the impact of Multiplex Connectivity Centres© on the lives of the people they are designed to help, it is important to appreciate that many candidates who pass through Multiplex Connectivity Centres© have suffered significant disadvantage and been out of full-time employment for many years. Indeed, many have never had a job before and have come from families which have suffered multi-generational unemployment.

Evidence collected using the above approach to social impact assessment shows that this has been a life-altering experience for many candidates, helping them overcome formidable cultural, economic, personal and social barriers to employment. In terms of specific impacts, this has not only given the candidates a new sense of purpose, self-confidence, pride, accomplishment, self-worth and dignity, but it has also developed a significant set of new skills, knowledge and attributes which will position participants strongly for sustained employment into the future.

The following statement made by an Indigenous candidate, who completed one of Multiplex's employment programmes, sums-up the social impact that they can produce:

> I came here to change my life and this has given me a chance to take a new direction, to have hope and to help me see how I can support my family and make them proud.

Finally, the social impact measurement process indicates that there have also been positive organisational outcomes for the construction industry partners involved. These include: leaving a positive legacy in the community; increased cultural awareness; increased staff satisfaction and happiness; personal growth; enhanced teamwork and leadership; innovation; demonstrable corporate citizenship; positive image and brand; mentoring skills development; workforce diversity; extra resources; staff motivation and engagement; satisfied clients and wider customer base; and improved cross-sector collaboration.

Summary

This chapter has described an innovative approach to creating social value by Multiplex which has been developed over many years of refinement and collaboration across a range of sectors. The challenges of forming effective cross-sector partnerships are many and require experience of working across those sectors and innovative ways of breaking down the many in-built institutional and cultural barriers to cross sector collaboration which exist. While governments are wanting innovative solutions to welfare provision through greater cross sector collaboration, making it happen in practice in much more challenging that it may first seem. Leadership and new skills are required, new relationships need to be formed and new incentives need to collaborate created. True collaboration requires a new way of thinking which requires system change and which relies on trust and strong inter-personal relationships at both an individual and organisational level which have been developed through an understanding of each other's needs and drivers. Multiplex's experience is that true cross-sector collaboration takes time and a great amount of energy, experience and commitment. Lessons have to be learnt along the way and approaches refined in response to the needs of each community in which one builds. Every construction project community is different. Not only will community needs differ but the organisations which will need to collaborate to meet these needs will also change, necessitating innovation, sensitivity and adaptability on every project.

9 Latch, Canopy, and CITU, Leeds, UK

Three unique SMEs providing secure housing

Ani Raiden, Martin Loosemore, Andrew King and Chris Gorse

In this chapter we introduce three small- to mediumsized enterprises (SMEs) whose operations focus on the housing sector. Social value is delivered on multiple levels, from changing individuals' lives through training and employment prospects and provision of secure housing, to regeneration and uplift of the local built environment and engaging communities.

Latch

Latch is a unique charitable organisation set up under the Cooperative and Community Benefit for Society Act 2014, as a limited company to serve the community. The key driver for Latch's management committee is to provide housing for people in difficult situations such that their circumstance would otherwise prevent them being able to secure housing. The company's operational terms describe their target group as 'necessitous', thus not having the necessities of life, in need and at risk of social exclusion. It is clear that, in most cities, some individuals and families, by circumstance or a chain of events, become difficult to house, have experienced social exclusion and their situation places them outside of common housing schemes and or employment. Such people need help to step back into the community and housing. Through housing, activities and volunteering, Latch have been able to make a positive change to the lives of many people and run the company as a going concern.

Background and more recent impact

Latch was established to help the homeless, vulnerable and those that have fallen out of work prospects. As a company, they started by refurbishing derelict and run-down houses in some of the more deprived areas of Leeds, a city in the North of England, United Kingdom. Through small grants, the help of volunteers and effective on-the-job training, their achievements, in terms of properties refurbished into decent homes standard, have grown steadily over the years. Latch has managed to transform buildings and lives by re-engaging its tenants with the local community and creating areas which now have a more prosperous outlook.

In 1989 Latch began working on derelict empty properties engaging homeless volunteers to create their own housing. The following year the company leased two properties from the local council. Since this time the organisation has engaged in self-build projects for young people, renovated flats and houses, provided a hands-on community training workshop, housed nearly 200 people, has 66 of its own homes and makes a positive impact on many disadvantaged groups. Latch have also piloted their own low energy homes, offering thermal upgrade services to homeowners, using staff that began in their volunteering scheme before progressing to employment.

Examples of those helped include:

- Homeless people without work prospects and situations that prevent them obtaining reliable references.
- Refugees with limited local language skills and unfamiliar with the requirements of running a house.
- Families that have previously been victims of social and mental domestic abuse, forced to leave their previous home into communal refuge, and without Latch's assistance may not have been able to transfer back to independent living.
- People without work experience gain valuable experience and find permanent work.

Latch's work goes beyond renovation and on-the-job training, it clearly recognises that individuals need to be socially connected to reengage in the community. Through Latch's activities, such as team days out, tenants and those that work with Latch build effective relationships, networks and develop confidence and self-worth enabling positive life changes to be achieved.

Developing, volunteering and working alongside experienced trades

Most of the renovation work is done by Latch staff and volunteers. Latch's staff and some of the volunteers have professional and trade experience, whilst others learn new skills as they work on site. Volunteer's engage for a variety of reasons, not simply gaining experience. The social benefits of the Latch community are significant and it is clear that those who engage and form part of the team are proud to be associated with the organisation and its achievements.

Latch's key aspects of social value and enterprise focus on work with vulnerable communities in order to:

- develop a quality product of value;
- allow individuals to benefit from volunteering and supporting with on the job skills development and training;
- enhance and improve the area;
- provide a positive outlook for those disengaged;
- help the most vulnerable to change their lives;

- create safe and comfortable environments;
- benefit all engaged through multiparty collaboration;
- share success stories;
- generally, move forward with new enterprise and ventures.

Through placing social engagement and valuing lives at the core of its enterprise, Latch has managed to change lives, improve individual's outlooks and provide an enterprise that is stimulating micro-economies in areas of need. Their business model is registered as a Community Benefit Society, a form of a charitable organisation, that is financed by donations and grant income.

Canopy: a community housing project

Canopy is a community housing company, founded in 1998 and based in inner city Leeds, which is one of the largest cities in the United Kingdom. Canopy is part of a 'self-help' network of community-led housing ventures that exist across the United Kingdom (Self Help Housing Org 2016). These organisations, their work, activities and impact are prominent in the larger inner cities and other disaffected or deprived areas of the country. They offer points of communication to engage with communities that need support. Many of the projects link in with organisations in the built environment, including contractors, designers, planners and trades, and as part of their programme of work offer an established channel of contract for socially aware professionals to engage with.

Canopy renovates empty and derelict houses and transforms them into homes. Their projects call for expertise, products and services from across the field of construction and rely on reliable and skilled experts to assist with voluntary work or apprentice schemes. There are many ways in which construction organisations can get involved with these self-help groups; one of the more common ways is to engage with the voluntary aspects of the self-help groups and offer work experience or apprentices as a stepping-stone to future employment. The prospect of future employment, through the partnering organisations, is attractive to those that believe they have little hope of gaining real employment. The self-help schemes offer an opportunity to gain experience in addition to further opportunities beyond the self-help work. Once experience is gained through voluntary work or apprenticeships, future job prospects are increased. The voluntary workers gain valuable training, develop capability and competence in their work as well as self-management skills. Once experience is established, they can become attractive to employers in need of capable and trustworthy staff. Such schemes are helpful in developing confidence and enabling a more effective and clear step into employment.

This way, Canopy goes beyond simply renovating void properties, it also uses them as a catalyst for community change. The work brings volunteers from the local community to learn skills, increase confidence and break down barriers. The act of making changes to derelict properties prevents the properties becoming a place for vandalism or a dumping ground and instead transforms buildings into homes, which are provided by the community, thereby making significant improvements to local neighbourhoods.

Most of the upgrades and renovations that Canopy undertake are made possible through their industry and local authority partnerships, grant income and volunteering. Volunteers come together and help new tenants to paint, decorate and furnish their new homes. Alongside the renovation of void properties, training and apprenticeships, Canopy also supports:

- Housing cooperatives and projects
- Refugee support
- Community Enterprise

The community housing companies, like Canopy, are led by the community and so they offer a unique way of engaging directly with those that need help. Typical to these schemes are the following attributes:

- community led or high level of community engagement;
- tackling poverty;
- engage and educate general public to remove misconceptions about those that need help;
- volunteering and training schemes;
- site based vocational qualifications for volunteers;
- properties renovated to comfortable living standards;
- reduce and take people out of fuel poverty, through thermal upgrade schemes;
- help people manage their lives and home, financially and socially;
- members often invest their own time in their homes and the houses of others in the scheme;
- reuse and recycle wherever possible;
- make effective use of local grants to start new initiatives;
- engaging the local community in community enterprise.

Similarly to Latch, Canopy's business model is registered as a Community Benefit Society, a form of a charitable organisation, that is financed by donations and grant income.

CITU: eco-developers with a strong focus on social integration

CITU's ethos is to offer a model of how to build low or zero carbon neighborhoods that provide healthier, smarter and better-connected communities and cities. Certainly in Yorkshire, the scale and type of projects that CITU develop make them rather unique. They have a reputation for working within deprived communities and developing the areas through master planning with the councils and other groups, working together to create strong socially integrated areas with good education and economic prospects. The first major project that CITU undertook was to redevelop and refurbish a derelict tower block in Beeston, Leeds, UK – The

Greenhouse. At the time of the development, the area had a high proportion of unoccupied and derelict properties. Deprivation and problems with health and wellbeing in the area were high, making it suitable for local and international grants that aim to tackle poverty.

The Greenhouse is an eight-story, mixed-use block of flats. The building is supported with wind power, ground source heat pumps, a communal heating system as well as other technology that bring energy efficiency. The building is mixed use, and has a combination of rented and owned property. When other developers were suffering with reduced sales of properties through the recent recession, CITU were able to continue selling and renting properties and introduced new business into the area around Greenhouse in Leeds. There are many aspects to the property and approach that challenge current convention, but the vital change is in the demographic that is helping to lift the socio-economic status of an area, bringing considerable social value to the local community.

CITU have also developed an area in Sheffield that was previously one of the oldest industrial sites in Sheffield. The Kelham site is an area that has been going through a great deal of urban regeneration, and CITU has taken a leading role. The previous industrial waste land, left as a result of the closure of mills, steel foundries and other heavy industries, is now the home to mixed-use development, with a bakery, restaurants and enterprise zones. With a strong emphasis on low to zero carbon and innovation, together with maintaining local heritage and bringing local employment through new community enterprise, the development is attracting a new demographic to the area, with a strong emphasis on community.

The most resent project that CITU have engaged with is the Climate Innovation District close to the centre of Leeds. It is the first ecological district of its scale in Europe, and again has the potential to bring positive change to a previously deprived area. The development is targeted to provide a catalyst for productive, sustainable and inclusive economic development. The masterplan for the development creates a step change for the city. Those that buy or rent one of the properties will take a stake in the future by being part of a not-for-profit organisation (Community Interest Company – CIC) that owns and controls the development. Any funds generated stay within the development and all of the households have a vote on the future investment decisions. The residents pay an initial bond, which is equivalent to the freehold and also gain control and ownership through the CIC. All of the on-site renewables, providing electricity, water and data costs are owned by the CIC. The residents gain access to utilities at cheaper rates than for-profit energy companies and generate revenue by selling excess solar energy. The development is in a deprived area of Leeds, close to other housing areas which have some of the worst poverty, education and health demographics in the country. The masterplan also includes a vision for a school, which together with local training and education facilities aim to achieve and high quality inner-city environment with greater diversity and social integration. The scheme goes well beyond what is required of the developer, and working with the council, CITU is managing to add a considerable social return to the area, lifting families out of poverty through access to facilities, education and cleaner environments.

CITU is a registered company, and so their company business model is financed by the property sales and rental income. Each CITU development is also self-sustaining through the surpluses generated by the not-for-profit Community Interest Companies attached to the project. These ensure the long-term viability and stability of the developments, and a sustainable future.

Summary

The three organisations profiled in this chapter, Latch, Canopy, and CITU, focus on creating social value in the housing sector by changing individuals' lives through training and employment prospects, and provision of secure housing, whilst at the same time helping to regenerate and uplift the local built environment. Latch and Canopy do this through a registered Community Benefit Society business model, a form of a charitable organisation, that is financed by donations and grant income. CITU operates as a registered company, and sets up self-sustaining not-for-profit Community Interest Companies with the view of securing long-term viability of its eco-developments.

Reference

Self Help Housing Org (2016) *Self Help Housing Org: Promoting community driven housing initiatives*. Available online at: http://self-help-housing.org/case-studies/case-study-2/.

10 Deconstructing social value in decommissioning

Industrial heritage at Dounreay, UK

Cara Mulholland, Paul W. Chan and Kate Canning

In this chapter, we analyse a case decommissioning the nuclear experimental facility at Dounreay. This is one of 17 sites being decommissioned by the Nuclear Decommissioning Authority (NDA) in the UK, and the total programme comprises projects that will last over the next 120 years at an estimated cost of around £119bn; the decommissioning of Dounreay is estimated to cost just under £3bn (NDA, 2017).

We first outline current developments of the decommissioning of the Dounreay experimental site and facility (herein after known as 'Dounreay'). We review current concerns about the site closure and plans for its afterlife and the site remediation. We then go further into the history of Dounreay to argue that current developments are not new. By tracing how residents around Dounreay have coped with historical changes in the past, we identify parallels in the discussion of what matters from a social value standpoint, then and now. Finally, we take a glance into possible futures for Dounreay and argue that social value is more than just a set of quantitative metrics to show the impacts of work done on stakeholders. Social value is in constant flux as stakeholders create and improvise narratives of what they consider to be important.

Dounreay: from ushering a brave new world of energy security to securitising the site

> If Dounreay hadn't been here, and assuming no other big thing was brought in, Caithness and North Sutherland would not have what it has today.
>
> (Site manager)

Rurally situated on the north coast of the Scottish Highlands in the county of Caithness, amongst a flat open landscape with very low-level vegetation, Dounreay as it stands today is a result of fluctuating strategies, timescales, and structures. The large sphere which housed one reactor is an iconic reminder of the significant nuclear scientific research which happened on the site, as seen in Figure 10.1. The site is surrounded by mostly farmland with a number of rare species of flora and fauna onsite and around and a stream that flows across the site boundaries. There is an adjacent beach and castle to the north of the site and the neighbouring Ministry of

Figure 10.1 Photograph of the Dounreay site showing the iconic 'Sphere'. July 2017[1]

Defence site to the west (Magnox 2014). A railway connects the area with the south of Scotland, and a well-maintained A-road runs along the east coast, though this is often blocked by slow moving traffic without passing points. There is also an international airport that services a few flights connecting to Aberdeen and Edinburgh in Scotland.

The fast breeder reactor programme and UK energy security

The creation of the Dounreay site can be traced back to the 1950s when the UK government developed interest in fast breeder reactors to meet its long-term energy needs (NDA 2011). Due to the pioneering nature of the technology, which meant working in very uncertain conditions when compared to industry norms at that time, there was a need to find a site remote enough to safely accommodate the fast breeder reactor. The Dounreay site was considered an appropriate choice partly because of its geographic location and partly because it was an area with a declining population and a lack of investment and employment opportunities (Hill 2013). By 1962, it was hailed as a success for being the world's first fast reactor to supply energy to a national grid, even though the output generated was often suboptimally below its full capacity (Lehtonen and Lieu 2011; Hill 2013).

Throughout the 1960s and 1970s, debates ensued around the efficiency of the Dounreay fast breeder reactor, safety concerns resulting from periodic leaks, and the desire to maintain Britain's lead position in nuclear energy technology in a context of growing diversification of its energy sources (see for example Dalyell 1977). After much deliberation, the UK government decided in the mid-1970s to further invest in the fast breeder reactor programme. However, support of Dounreay declined following an explosion in an onsite waste shaft in 1977 and discovery of radioactive particles on a nearby beach in 1984. The decision to invest in the breeder programme was later reversed by the Conservative government in 1988, in part due to these safety concerns (and also post-Chernobyl) and other political factors that saw a policy shift away from dependency on nuclear and coal technologies for supplying Britain's energy mix. In a pre-social value world and with a sudden announcement of closure there were no plans in place.

With a lack of clarity on the site strategy, the initial closure announcement created a period of uncertainty. Job losses were inevitable, and from 1988 to 1993 the number of employees fell from 13,600 to 8,300. This led to a breakdown in trust. People on site were unsure of what they were going to do after the initial experience of job losses. An ex-employee who worked on several complex projects over ten years commented recently that 'I don't think they'll ever close it down', as they saw individual projects with changing goals and end dates.

Securitising the site: making sense of uncertainty and change

Issues of health and safety and security at Dounreay led to a subsequent increase in public scrutiny, which in turn brought about a number of major changes. As well as an ongoing issue with radiation contamination of the local seabed, in 1998 a digger accident caused power loss to critical structures in the fuel cycle area. This accident raised public concern over the use of private contractors in managing nuclear safety, which in turn resulted in the decision to stop contracting in lead safety officials. Later that year the Health and Safety Executive (1998) undertook a safety audit identifying 143 recommendations to address security and safety weaknesses.

A response to these calls for safety reforms can be seen in Dounreay through the nuclear safety flyers distributed to the local community since that time (see Figure 10.2). By rhetorically framing these concerns as 'nuclear emergencies', Dounreay established a sense of urgency. Furthermore, by emphasising 'community policing', the intention was to empower local communities to take control. By 2005, spaces and forums were created, for instance, on Thurso High Street where stakeholder committees would regularly meet with the facilitation of the Dounreay engagement office.

Dounreay also responded to the Health and Safety Executive audit of 1998 by planning and driving programmes for the decommissioning work. In 2000 the Dounreay Decommissioning Plan was set at 60 years at a cost of £4.3 billion; by 2007, this timeframe was brought forward to 2032 with a forecast cost of £2.9 billion and in 2008 this was again brought forward to 2025 at a cost of £2.6 billion. In 2018 the programme is still set for delivery by the mid-2020s at a cost of £2.9 billion. Despite concerns over securing the site, the pioneering technology on such a large scale at Dounreay had been constructed without deconstruction in mind. Nuclear decommissioning challenges from fast breeder reactors had not been faced or solved elsewhere before, and the experimental nature of many of its redundant facilities meant that the clean-up and demolition would require

Figure 10.2 Rhetorical framing of nuclear securities. July 2017

innovation as well as great care. Thus, Dounreay was confronted by the paradox of needing to address safety concerns by speeding up the decommissioning process while ensuring that the first-of-a-kind technology was handled with care. With such complex problems to solve, planning for uncertainty has proven difficult, with the programme of work and timescales still seemingly a moving target.

The site went through yet another major organisational and managerial restructuring in 2006. This came after the NDA was formed in 2005 with the UK Energy Act (2004), and the Dounreay Site Licence Restoration Ltd (DSRL) was established in 2008 to take over from the UK Atomic Energy Authority as the site licenced company. This restructure began the process of thinking not just about decommissioning the facility at Dounreay, but also planning for site remediation to accommodate life after decommissioning. As one site manager noted,

> The mission is very clear. Since 2006 really when we had that second sort of re-shake, it became very clear that the only message in town was to decommission and turn the site back into a brownfield site.

In contrast to the decade of uncertainty and job losses following the initial announcement of closure, gaining clarity of the mission for decommissioning and site remediation meant that the employees and local community started to re-engage in earnest with projects with a longer vision in mind. Decommissioning and remediation work on the 20-hectare site was planned to take place until the mid-2020s, with a further period of institutional control that will see a final end-state and handover in the 2030s. Twelve hectares onsite have also been designated for a low level waste repository which will be maintained by a small number of employees for up to 300 years (Magnox 2014).

Finding a focus: dawn of the decade of planning through socio-economic strategies

With the re-energising of the local community, Dounreay saw a period of formalised planning through the process of developing socio-economic strategies and documenting impacts of investment alongside the technical work on-site. The first Dounreay Socio-Economic Development Plan was published in 2008–2009 as an immediate response to the NDA's first socio-economic policy report (NDA 2008). These reports were updated annually to outline progress made on an eclectic range of projects, including business development, community projects, infrastructure, and arts and culture. Other strategy documents included the Heritage Strategy of Dounreay (DSRL 2010) which focused on how historical objects, photographs and other paraphernalia stored in the Caithness Horizons Museum in Thurso can be used to revive the cultural heritage of the local community.

Thus, the capture of social value in the Dounreay case pre-dates the introduction of the Public Services (Social Value) Act in 2012. Indeed, the 2012 Act led to similar strategic lines taken, including socio-economic outcomes, heritage, health and safety, environmental impacts and, financial and technical objectives (see NDA 2016a). By 2015/16, the Dounreay Socio-Economic Alliance (2016) recorded £90 million spent on wages, and £10.82 million of the £139 million spent on the supply chain could be traced to local SMEs.

A nuclear vision for the north of Scotland

The transformation sketched out in the preceding sections illustrates longitudinally the ongoing process of how Dounreay and the local community have coped, more tactically than strategically, with the announcement of the closure of the facility over 20 years ago. Such disruptive change is certainly not new, as we shall see next where we reflect upon the time before Dounreay came into being, and to show how questions about social value were pertinent back then even though the term was not even invented yet.

Dounreay, in the county of Caithness situated in the Scottish Highlands, is somewhat on the periphery of mainland UK life. The historic remains of Caithness have been left marked into its landscape. The Iron Age brochs, cairns

Figure 10.3 Ruins of religious landmarks dating back to the 13th century. Thurso, July 2017

and standing stones structures are scattered throughout the county. There are sur-
viving religious and agricultural structures dating back hundreds of years, as seen
in Figure 10.3. The land is still divided up into smallholdings known as crofts,
which hold the history of landlords and harsh treatment of the native Scots, with
many of the current Dounreay workers still maintaining their small agricultural
patches.

This rural location was once known for agricultural and fishing industries
stretching back hundreds of years. Hemp was widely farmed to produce quality
fishing nets, and Wick in the East of Caithness became the busiest herring port
in Europe in the 1800s. After peaking in the early 1900s, new fishing methods
and the loss of fishermen to the navy in the First World War meant a decline in
the industry. This decline continued through the Second World War, with many
small enterprises not able to compete with growing markets and intensive fishing
methods after:

> There was fishing and farming, and of course those were heading towards
> decline. So, this was a big industry to bring to the area. And it was a big
> exciting vision for the future. It was cutting edge research, and in some ways
> it still is today.
>
> (Site manager)

In 1940 Caithness suffered the first daytime bombing raid on mainland UK in
the Second World War. As a response, military defences and bases were established
in the county, including the airfield where Dounreay now sits, seen in Figure 10.4.
Once the war was over the UK Government turned towards nuclear research as a

Figure 10.4 Former runway of the RAF airfield, now turned into a car park for the Dounreay nuclear research facility. Dounreay, July 2017

peaceful pursuit, wanting to harness the energy potential. The facility was born as a response to the UK's move toward nuclear science for energy.

As the newly-formed UK Atomic Energy Agency looked for somewhere to locate their new nuclear research facilities, an isolated place close to water was ideal. The community in Caithness was also looking for job opportunities, and so the nuclear vision was brought to Dounreay along with all the associated opportunities for growth. The disused brownfield site attracted rural development, with markings of the airfield still found at Dounreay. Costing a total of £28 million and taking only three years to build, deconstruction and decommissioning at the end of the facility's life were not considered. This led to the legacy of technical challenges:

> They built it here because it was remote, and nobody lived here. You know, it was the furthest place as far away from London as possible.
>
> (Former representative of the trade union)

And so arrived the 'Atomics', a colloquial term used (at times in a derogatory manner to differentiate between the outsiders and the locals) to refer to the leading scientists, engineers and technicians who moved to the area with their families to work in the research facility. This increased the population of a declining community by around 17 percent between 1951 and 1961, compared to an overall population growth in the UK of around 2.5 percent (Smith 1988). This in turn led also to new opportunities that encouraged the locals to stay:

> So, I think Dounreay was good. Not only in that it brought people up here, but it actually prevented a mass exodus of people from Caithness. Because until Dounreay came, people left, because there was no employment.
>
> (Member of the local clergy)

One of the site managers interviewed also recounted how

> [UK Atomic Energy Agency] built a lot of houses in Thurso, a school, new high school, technical college, hospital facilities. So much that we, that people take for granted today was born out of the need to support the mission here [...] So the whole social infrastructure, especially in the West of Caithness, was driven by what was happening here.
>
> (Site manager)

Dounreay provided onsite benefits such as dentists, banks and hairdressers. Houses, leisure centres, social clubs and local infrastructure were built to support the facility's incomers, incentivising additional investment in the area. Employment, education and training created a skilled community who had more opportunities available to them. As the Atomics moved to work in the new facility, their accompanying families not only became integrated into the community but also started to create social clubs in what was otherwise an isolated location.

The new arrivals, nevertheless, produced some tensions and contradictions. On the one hand, the creation of the Dounreay facility attracted highly-educated professionals into the area. As a member of the local clergy remarked, 'think of the gene pool, those people who I've said were leading [Dounreay], their children were highly intelligent'. In fact, Thurso had at one time the best high school education results in the UK. On the other hand, there were segments of the local community who were left behind. Thurso had the highest rate of teen pregnancies. It was a 'cosmopolitan' place to live if you were part of the highly-skilled professional elite, but a rather isolated place if you were not. The socio-economically deprived had few job opportunities if they were not directly employed by Dounreay. In contrast, Dounreay employees had access to educational and training opportunities. Many developed skills that then armed them for lifelong careers with continued professional development. Some used their skills to move into supporting supply chain companies, while some others used their newly-learnt skills to enable them to move elsewhere for employment.

As we can see from the time before Dounreay came into being, social value as we know today was not framed in terms of devising a plan and measuring the outcomes of that plan at that time. Rather, a technological need combined with addressing a societal problem created a moment of opportunity for a community to grow. Yet, this growth was not always uniformly positive.

Creating cultural heritage and legacy in a period of decline

In the period of 2010–2016 Thurso and Wick saw their populations decline. At 21 percent, the Highlands and Islands population had a higher percentage of over-65s in 2010 than the Scottish average of 16.8 percent. While Caithness had marginally more young people, it also has an aging population with a faster rate of increase in the number of older people than the rest of the wider region, as more young people move elsewhere for employment opportunities (Grangeston 2012).

Therefore, the region returns yet again to a period of decline. For some though, nuclear decommissioning at Dounreay is still seen as an important source of employment opportunity. As a member of the Archive staff remarked:

> They see decommissioning as an opportunity not something that's coming to an end. They see it as providing jobs.

Nevertheless, the socio-economic impacts of losing a major employer in an area is widely felt, with fears that a reduction in employment opportunities would lead to a vicious cycle of reductions in such public services as healthcare, education and transport, which in turn discourages new incomers or returners to the area. The moment of realisation of such downward spiral was best captured by the comments of a former representative of a Trade Union:

> And I think even though we know that we're all working ourselves out of a job and at some point that the numbers are going to reduce, for all of a sudden, we're going to lose 200 [...] we need to get rid of 200 [...] it was like, oh!

Moreover, there is a growing concern about the loss of skills and knowledge from previous time-served employees as they leave and retire. Dounreay was pioneering in nuclear education and training and there is a real risk of losing all that tacit knowledge (NDA 2016b). A Remediation Manager bemoaned:

> It's a worry that because nuclear is not such a big thing as it was in the past they're losing a lot of skills. Like [Dounreay] was a reactor, this was a centre of research. Those skills, there's a real danger these will be lost.

In an attempt to maintain the legacy of Dounreay, the NDA opened Nucleus, a new building to house the Nuclear and Caithness Archives, in 2017 (see Figure 10.5). Over a third of a million photographs and 200 tonnes of documentation about

Figure 10.5 The new Nucleus building. Wick, July 2017

Dounreay have now moved to the Nucleus Archives. The building also won a number of awards, including the recent Royal Incorporation of Architects in Scotland (RIAS) Award in June 2018. What is noteworthy is the summary of the judges' deliberation on Nucleus' entry (RIAS 2018):

> The judges unanimously felt that this was an impressive industrial project which has created an ethereal and beautifully sculpted building. [...] The relationship of the structure to the historic context of the site and the adjacent wartime airfield, together with the references to the Caithness context of lochs and a difficult climate, are beautifully articulated.

Thus, it is clear how this building was designed to symbolically blend the various representations of Dounreay and the region over time.

The future strategies for Dounreay and the surrounding Caithness region reflect both potentials of industrial and cultural heritage as a role in creating new industrial and culture-based jobs. One emerging area of job creation is in the technology and engineering field, with spin-off businesses from Dounreay and other renewable energy technologies (as seen at Wick Harbour in Figure 10.6).

Ironically, where Dounreay re-energised what was a declining agricultural and fishing community in the 1950s, agricultural and fishing heritage is now returning and thriving in Dounreay's demise. With a rise in tourism and growing passion and nostalgia for traditional skills, entrepreneurs are now seizing the opportunity for ecotourism and glamping on the vast open land. Some of this is, of course, invested by the NDA. For example, the NDA has supported the

Figure 10.6 New offshore wind energy structures ready to leave Wick Harbour. Wick, July 2017

redevelopment and growth of both nearby Srabster and Wick harbours with £2 million and £250,000 respectively, allowing Caithness communities to return to the sea, with growth in fishing and other leisure activities. None the less, it is worth noting that many of the heritage revival initiatives are a consequence of the drive and efforts of local enthusiasts. For example, public consultations in the early days when the closure was first announced unsurprisingly indicated that the public – many of whom were employees in Dounreay – wanted a new nuclear facility, thus preserving their way of life. Then there were ongoing debates over whether the brownfield site should be retained or whether Dounreay should return to green spaces. Whilst the financial costs and benefits of removing everything and returning to greenspace ultimately showed that this option was not feasible, the identification of options was not always straightforward. Engaging with any community of diverse views and needs is bound to be fraught with challenges. Yet, local initiatives continue to emerge.

As Arthur (1998) concluded, reflecting the strong feelings of some on the importance of the site's industrial heritage and asserting the future of the sphere:[2]

> That [Sphere] will remain: it has been listed by Scottish Heritage. The rest of the plant will continue to process nuclear material, at least until 2006. In 100 years or so, its doors will finally close. The beast is dead – but it will probably survive longer than anyone who killed it off.

Social value as an organic or a planned process?

The thick longitudinal description of the ebb and flow of developments in Dounreay demonstrates that what 'society' values is an organic process and outcome. Thus, the revival of a declining subsistence community with the advent of the fast breeder technology produced a virtuous cycle of new opportunities and services for the local community. This was not borne out of some altruistic social value strategy, but a drive to become a world leader in nuclear energy technologies coupled with the experimentation of pioneering, uncertain and unsafe technology. The result is that the experimental fastbreeder technology did indeed put Dounreay on the world map as a 'successful' international centre of nuclear energy research.

That is not to say there is no value in taking a planned, managerial approach to social value. As policy changes unfolded in the Dounreay story, the introduction of a clear mission of decommissioning and site remediation meant that the NDA could take back control by focusing on the measurement and management of socio-economic impacts of the site.

Today, the NDA takes a more open approach to public consultation, preferring to provide alternatives rather than to plan for the 'best' option. Given the long gestation periods of nuclear decommissioning programmes, this is sensible since the long timeframes will inherently bring about greater levels of uncertainty. The narrative of Dounreay shows that even if social value were calculated in the 1950s, unforeseeable events would have altered such 'value' many times over the past 60 years. What was at play in the Dounreay narrative is constant negotiation with a community that in itself changed over time with the entry of the Atomics. Thus, taking a longitudinal and more processual stance, social value could be seen as an ongoing process of replying to the changing contexts over time (see Chan 2016).

Nuclear legacy as industrial heritage

Dounreay is one megaproject within a mega-programme of decommissioning projects managed by the NDA. The estate is often referred to as 'the UK's nuclear legacy'. Such legacy is imprinted, in part, in the physical structures that are being decommissioned. But, as we have seen, the legacy is also found in the technologies, technological history, and the challenges of cleaning up the first-of-a-kind pioneering technology. Alongside these are the unseen effects such as tacit knowledge and cultural change. The legacy also extends to the land and the sea, flora and fauna of the place. These include both the countable and uncountable aspects of a sense of place. The RAF airfield, along with the fishing and nuclear heritage are masterfully blended in the symbolic representation of the Nucleus Archives and, as Historic Scotland (2015) would put it, the turning of industrial heritage into a destination choice for tourism and education so that marginalised and isolated communities can continue to flourish (see also the EU European Regional Development Fund 2014).

Summary

By studying the dynamics of social value creation within the megaproject context of Dounreay, we have opened up the concept of social value beyond the project boundaries. Instead of following prevailing approaches to quantifying social value, our point of departure is to produce more qualitative understandings of the changing character of social value. In the case of Dounreay, the themes of social value which arise are not simply emerging as a result of site remediation, closure and hand-over; some issues concerned the locals before the nuclear facilities were brought to site and throughout the Dounreay journey. This journey is far from a linear process; rather, the journey appears to be a circular, dynamic process of value creation (and in instances, value destruction and value re-creation). The narrative has seen several transformations, from the resuscitation of a declining subsistence town, to the decline of a nuclear experimental facility, to the revival of 'traditional' subsistence skills.

The Dounreay case presented in this chapter has shed light on the significance of temporalities – both in terms of timescales and time frames (periods) – in framing social value. We demonstrate the need for social valuers to move beyond the fixation on measuring value at a point in time to pay more attention to the shifting contexts of what society finds valuable over time.

Postscript: a brief word about methods

To piece together this relatively thick description of the historical and current developments at Dounreay, we adopted a qualitative case study methodology (Eisenhardt and Graebner 2007). This was informed by semi-structured interviews with nine participants from the site and local community, participant observational data collected across two site stakeholder meetings and four site visits (to the decommissioning site, nuclear history museum, local historical museum, and the Nucleus Archives), photographs and informal conversations captured in field memos, as well as documentary analysis of the grey literature. The data collection and analytical process centred around how the local community evolved over time as a consequence of the Dounreay site, both in terms of its creation as well as its closure, and how ideas of economic, social and environmental wellbeing were conceptualised, discussed and enacted.

Notes

1 Photographs were taken by the first author who undertook the fieldwork.
2 It is worth noting that while there were many in the local community who advocated for retaining and listing the 'Sphere' as a heritage structure to remind future generations of the nuclear legacy of the Fast Breeder Programme, the 'Sphere' was never listed. Contrary to the voices of the local community, the 'Sphere' is now scheduled for demolition.

References

Arthur, C. (1998) The experiment ends, but Dounreay lives on. *Independent*, 6 June. Available online at: https://tinyurl.com/Dounreayliveson [accessed 19th June, 2018].
Chan, P.W. (2016) Expert knowledge in the making: using a processual lens to examine expertise in construction. *Construction Management and Economics*, 34(7–8): 471–483.
Dalyell, T. (1977) Westminster scene. *New Scientist*, 73(1037): 280–281.
Dounreay Socio-Economic Alliance (2016) *Dounreay Socio-Economic Review 2015–2016*.
DSRL Ltd. (2010) *Dounreay Heritage Strategy*.
Eisenhardt, K.M. and Graebner, M.E. (2007) Theory building from cases: opportunities and challenges. *Academy of Management Journal*, 50(1): 25–32.
EU European Regional Development Fund (2014) *European Strategy for Promotion of Industrial Heritage*.
Grangeston (2012) *The Socio-Economic Impacts of Dounreay Decommissioning*. Available online at: file:///C:/Users/hmd3raideab/Downloads/Socio-economic-Impacts-of-Dounreay-Decommissioning-Final-Report.pdf.
Health and Safety Executive (1998) *Safety Audit of Dounreay*. Suffolk: HSE Books.
Historic Scotland (2015) *An Industrial Heritage Strategy for Scotland (draft report)*.
Hill, C.N. (2013) *An Atomic Empire: A Technical History of the Rise and Fall British Atomic Energy Programme*. London: Imperial College Press.
Lehtonen, M. and Lieu, J. (2011) *The Rise and Fall of the Fast Breeder Reactor Technology in the UK: Between Engineering 'Dreams' and Economic 'Realities'?* 3 October 2011. Science and Technology Policy Research (SPRU) and The Sussex Energy Group. Brighton: University of Sussex.

Magnox (2014) Dounreay Site Strategic Environmental Assessment Site Specific Baseline.

NDA (2011) Exotic Fuels – Dounreay Fast Reactor (DFR) Breeder. *July*. Nuclear Decommissioning Authority. Available online at: https://tinyurl.com/DounreayFBR Options-Jul2011 [accessed 19th June, 2018].

NDA (2017) *Nuclear Provision: The Cost of Cleaning Up Britain's Historic Nuclear Sites*. Nuclear Decommissioning Authority. Available online at: www.tinyurl.com/ UKnucleardecommissioning-Jul17 [accessed 19th June, 2018].

NDA (2008) *NDA Socio-Economic Policy*. Cumbria: Nuclear Decommissioning Authority.

NDA (2016a) *The NDA Value Framework*. Cumbria: Nuclear Decommissioning Authority.

NDA (2016b) *NDA Strategy April 2016*. Cumbria: Nuclear Decommissioning Authority.

RIAS (2018) The Royal Incorporation of Architects in Scotland Awards 2018. Available online at: www.ribaj.com/buildings/rias-awards-2018-scotland [accessed 22nd June, 2018].

Smith, J.S. (1988) *The Third Statistical Account of Scotland. The County of Caithness, vol 19*. Edinburgh: Scottish Academic Press.

11 Conclusion

*Ani Raiden, Martin Loosemore, Andrew King
and Chris Gorse*

The purpose of the book was to explore the emerging concept of social value in the context of the built environment by creating a conceptual and practical foundation to advance thinking and practice for every professional and firm involved with the entire life-cycle of construction from planning through design, construction, operations and facilities management. We have shown how built environment has a major impact on the communities in which it builds and that there is a mutuality of interests in doing so, although the skills and knowledge needed to manage the social dimensions of this impact have been relatively neglected in comparison to economic and environmental dimensions. To work towards filling this gap in our knowledge and practice base around social value, we have clarified the confusion surrounding what the term means and what the process of creating social value involves.

In the context of the built environment, our definition of social value is simply the 'social impact' any organisation working in the built environment makes to the lives of internal and external stakeholders affected by its activities, including those working in the industry and in the communities in which it operates. In particular, we have pointed to the need for the construction industry to develop new roles, networks, relationships, skills and knowledge to work across sectors. It is clear that new expectations to create social value through the emergence of social procurement initiatives which are themselves driven by wider trends in CSR and new public governance, raise many new challenges, but also many opportunities, for professionals and businesses operating in the built environment.

Within this context, we have advocated a systems approach to creating social value that recognises the need to take a holistic view of the many private, public, charitable, not-for-profit and community organisations and individuals involved in delivering social value to the community. We have also emphasised the need to understand the ethical and political context in which the social value is created. The many people involved in the creation of social value will see it and define it in different ways and the variety of perceptions will need to be managed sensitively since trade-offs will be necessary between the competing needs of many stakeholders. Furthermore, opportunities to create social value change over time in response to the social, political, economic and cultural environment in which a judgement about value is being made. It is also important to be aware that the idea of involving the private sector in tackling social problems is inherently political and highly controversial in itself. Finally, the lack of agreement and discipline around the practice of measuring and communicating social value ensures that there is little empirical evidence to support (or deny) the many claims that these relatively new approaches to creating social value offer over traditional government welfare systems. So arguments for and against involving private firms in the creation of social value can easily be manipulated one way or the other.

Despite the inevitable politics and the many barriers to social value creation in the highly commercial built environment and construction industries, we argue that the timing for someone to 'step-up' and show some leadership in this area is perfect. Not only is there a growing social need, but there is an unprecedented global building and infrastructure pipeline to leverage, and a deepening skills shortage that can be addressed using social procurement to access a more diverse workforce including women, people with disabilities, and ethnic groups who have been traditionally excluded from work in the built environment. We have advanced the beginnings of a valuable framework for readers to judge the social value proposition of their businesses and projects, to identify what forms those social value propositions may take and develop some strategies to enable firms to maximise that value. We have also provided numerous practical examples and evidence of how social value can be created at every stage of the building life-cycle from inception, through urban planning and design, to individual building design, construction and facilities management. It has become clear that the creation of social value requires cross sector collaboration within the built environment as well as outside it. In other words, the maximisation of social value in the built environment depends on an integrated approach which requires the coordination and collaboration of every profession in the building procurement chain.

The legal and ethical framework

As we point out in Chapter 2, the emerging legal framework around social value has provided an important imperative and incentive to consider social value creation as a new form of practice and an element of business strategy in the built environment. This framework varies in focus and approach from country-to-country; being targeted and prescriptive in some countries and voluntary and non-prescriptive in others. However, regardless of context and approach, these frameworks place new duties on professionals in the built environment to consider social value which will trickle down the supply chain and affect every firm in the industry, no matter how large or small. The new requirement to consider how organisational and project specific activities create or destroy social value is not going away. The main lesson to emerge from our discussions around the legalities of social value creation is that social value is not something to arbitrarily bolt on to the back of a contract. Rather, the maximisation of social value from a legal perspective should involve rethinking a contract and how it is procured from the very start. A strategic approach to creating social value is needed and leadership and effective project management are essential. Setting social value objectives at the start of a project is critical so that tender/contract requirements are a core part of the contract deliverables and are stated in ways which are realistic, relevant, measurable and capable of being monitored and verified. This means that social value requirements must be linked to the subject matter of the contract so that the social value requirement is demonstrably a critical element of what the employer is seeking to achieve through that contract. While this can be a rather obscure requirement, this is best done by ensuring that the social value requirements included in a contract relate to the heart of an employer's or client's social value policy and to those objectives which are core or critical to successful contract delivery. Linking contract clauses and deliverables to a social value policy enables employers to achieve a consistent approach to social value, linking it to social value objectives for an organisation and ways in which it will seek to implement them. In other words, social value is not something one creates overnight. Social value creation on individual built environment projects is just one part of a wider strategy and set of values designed to create social value across an entire organisation.

Another key message from this book is the notion that social value as a concept is inherently grounded in notions of right and wrong which means that an understanding of ethics is foundational knowledge for anyone wishing to come to grips with this area. In the first instance there is a need to understand that there are both moral and non-moral reasons for creating social value. Doing the right thing because it is the right thing is distinctly different to performing the same action, with the same consequences, but for other motivations such as compliance with laws or contracts. Drawing on various ethical schools of thought we show just how easy it is for decision-makers to construct an ethical argument for-or-against any new idea and how ethical codes can clash in making value judgements about a social programme/ initiative. We also show that while useful, even a deep understanding of ethics is not likely to provide a definitive judgement of social value, but only serves to provide us with a set of general principles that we can use to develop our own position.

Co-creating social value

In terms of creating social value, one important lesson from this book is that social value is co-created at the intersection of public, social and private spheres of economic activity which highlights the need to form new inter-organisational partnerships that bring together governmental, business and non-profit domains. We also point to the changing nature of supply chains in the built environment, and the rise of hybrid organisations like social enterprises which pursue a blended approach to value creation and seek to balance both business and community interests in sustainable business models. These organisations are increasingly partnering with large national and multinational organisations that have tended to be commercially and economically driven, offering very few social benefits and often taking work away from local businesses and communities. However, there remain many barriers to the successful integration of social enterprises into what is still a highly commercial and incestuous construction industry with strong relationships and path dependencies which are notoriously difficult to break.

We also point to the importance of employment and training as some of the key pillars of creating social value through construction work. Construction is one the world's largest employers and is unique in that it operates in our most disadvantaged communities and has low barriers to entry for many people suffering disadvantage. Employment and training opportunities provide an enormous number of economic, mental and physical health, social and cultural benefits for people and is a powerful force for building more prosperous and resilient communities. Creative solutions which open the built environment to those who are normally excluded from its labour markets not only enables firms to create social value but also diversifies the workforce producing many advantages such as greater innovation and productivity, although news skills need to be developed to integrate and effectively manage these new cohorts.

Measuring social value

In Chapter 5 on social value assessment we argue that social impact measurement evaluations should never be taken as precise, although increasingly techniques such as SROI provide the illusion of accuracy to those who do not understand it's inherent and widely recognised limitations. In assessing social value we point to a wide range of opportunities for discretion in the evaluation process which can bias results, from who carries it out, to the selection and identification of indicators, to deciding which stakeholders to consult and involve, to deciding what data is collected and by which methods, and finally to the analysis and presentation of results where there are often strong incentives for organisations to inflate impacts or to be selective in presenting their results. Nevertheless, we recognise that being able to assess social value is crucial to enable managers to make more informed decisions, to identify where the greatest social value is being created in a business or project and in its supply chain, and to communicating effectively with key stakeholders such as staff, customers, funders, investors, communities. To this end, and while we recognise that the field of social value measurement is young and plagued by many unresolved issues such as inconsistent terminology and methodologies and questions around validity and reliability, we draw on a range of widely respected frameworks to present our own practical five step social value measurement approach which is rigorous yet simple and straight forward.

The first step in our framework recognises the importance of a bottom-up approach that places the community and its social needs and priorities at the foundation of any social programme and correspondingly any subsequent assessment of its impact in that community. So, the first step in any social value assessment should be to undertake a community needs assessment to understand community social needs, priorities, strengths and weaknesses. The second step is to develop a programme to meet these needs taking into account what needs are already being met and working collaboratively with other organisations in the community to do that. This programme should be underpinned by a theory of change that shows how the intended programme activities lead to the social outcomes which meet those community needs. The third stage of our social value assessment methodology involves identifying and prioritizing metrics for measuring whether the key social outcomes as identified by the theory of change have happened. It also involves establishing how these outcomes will be measured using valid and reliable metrics, methods of data collection and analysis in a way which minimises potential bias. This requires a deep understanding of research methods, and we argue that best social impact assessments use a combination of quantitative and qualitative techniques of data collection and analysis, and both primary and secondary data collected in a variety of ways and from a variety of stakeholders and sources. We also point to the important and often neglected question of ethics in social value measurement. We feel this is critically important given the need to collect data from vulnerable groups and the potential for the research process itself to do more harm than good if not conducted in a sensitive and responsible manner.

Having established the community needs and priorities, created a programme and its theory of change, and developed a reliable and valid measurement framework to collect and analyse social outcome data, the fourth step in our approach to social value measurement involves implementing the social programme and assessing its social impact. Like ethics, programme implementation is often missing from literature on assessing social impact, yet it is a crucial stage of the social value creation process since poor implementation can undermine any intended social impacts, making the process of assessment futile. It is only now, after all these steps that one is able to assess social value with any degree of reliability and rigor. The process of assessing social impact involves assessing whether the planned outcomes in the Theory of change are actually achieved in practice taking into account counterfactuals. Too often counterfactuals are not accounted for, and the results are grossly exaggerated claims about the impacts of the social value programmes on the community. Another trend which we discuss is the tendency to monetize social value outcomes by using techniques such as CBA and SROI. At this point in time, this remains a highly contested area and those who adopt such approaches need to exercise considerable care and thought since they open themselves to numerous criticisms of subjectivity, unreliable data, inconsistent methodologies, inbuilt biases and dubious assumptions underpinning such approaches. The final stage of our approach to assessing social value relates to reporting the results clearly and reliably to target audiences. To this end we suggest a number of good practices to follow and note that although sustainability reporting remains largely voluntary in end-of-year company reports in most countries, reports of social impact are increasingly incorporated into company annual reports, either as separate sustainability reports or as part of an integrated annual report covering triple bottom line of economic, social and environmental indicators.

The whole issue of social value assessment is arguably the most contested and hotly debated area in the field of social value and we end our book by re-iterating a critically important point – that the process of assessing social impact should not be an end in itself but a means to target strategy and change behaviours amongst stakeholders that can maximise positive outcomes for the communities in which we build.

Index

Milton Keynes UK
Ingram Content Group UK Ltd.
UKHW031148141024
449569UK00024B/977